Palgrave Studies in Anthropology of Sustainability

Series Editors

Marc Brightman
University College London
London, UK

Jerome Lewis
Department of Anthropology
University College London
London, UK

Our series aims to bring together research on the social, behavioral and cultural dimensions of sustainability: on local and global understandings of the concept and on lived practices around the world. It will publish studies focusing on ways of living, acting and thinking which claim to favor the local and global ecological systems of which we are part, and on which we depend for survival. Political pressure surrounding sustainable resource governance shapes regimes of measurement and control and the devolution of risk and responsibility. Scientific cultures of sustainability are generated out of concern over the need for 'green' technologies and materials. Popular discourses of scarcity of resources or capital increasingly lead to challenges to cosmopolitan and egalitarian ideals (human rights, the welfare state), fed by fears over the sustainability of social systems and civilizations in the face of global change. Meanwhile an array of social and cultural transformations are occurring that seek to offer ways to live (and produce, consume…) more sustainably. Calculations of sustainability raise questions of value – a vexed political affair. An anthropological approach will help understand these emerging phenomena.

More information about this series at
http://www.springer.com/series/14648

Roxana Moroşanu

An Ethnography of Household Energy Demand in the UK

Everyday Temporalities of Digital Media Usage

Roxana Moroşanu
Loughborough University
Loughborough, UK

Palgrave Studies in Anthropology of Sustainability
ISBN 978-1-137-59340-5 (hardcover)   ISBN 978-1-137-59341-2   (eBook)
ISBN 978-1-349-93474-4 (softcover)
DOI 10.1057/978-1-137-59341-2

Library of Congress Control Number: 2016942647

Cover illustration: © amana images inc. / Alamy Stock Photo

Printed on acid-free paper

This Palgrave Macmillan imprint is published by Springer Nature
The registered company is Nature America Inc. New York

To my parents, Dana and Cristi, and to my brother, Dragoş
To Ste, Lee, and Brett

ACKNOWLEDGEMENTS

I have benefited from the comments and criticisms of participants in numerous conference sessions and workshops where parts of this work were presented, and from more extended feedback from the people who kindly read (parts of) the manuscript as it developed: Alyssa Grossman, Val Mitchell, Daniel Miller, Line Nyhagen, Karen O'Reilly, Sarah Pink, Marilyn Strathern, and Palgrave's anonymous reader. I cannot thank them enough for their thoughtful and constructive reviews, and for their support.

I am immensely grateful to the people who took part in my research, for letting me inside their homes and inside their lives, for trusting me, for talking to me, for giving me cups of tea, and for letting me learn from them. More than making my research possible, meeting them was a life-changing experience.

I would like to thank all the members of the Low Effort Energy Demand Reduction (LEEDR) project, led by Richard Buswell, for welcoming me as part of the team and for working along with me for over four years. As part of the LEEDR project, the funding for my research came from the Engineering and Physical Sciences Research Council of the UK (Grant no. EP/I000267/1).

The members of staff in the Social Sciences Department at Loughborough University have been very welcoming and supporting, and we have had many stimulating conversations over the years.

I am grateful to my friends Zina and Richard; Christine, Vic, Helena and Humphrey; Val and Darryl; and Bob and Angie, who made me a

Morris dancer, took me with them all around the UK, and made me feel that I belong.

During the process of transforming my doctoral dissertation into a monograph, Steven Firth provided every kind of encouragement, together with patience and care. Thank you, from all my heart.

CONTENTS

NOTE ABOUT THE AUTHOR

I see myself as an optimist, and I regard the opportunity of engaging in long-term ethnographic research as providing immense personal gain by bringing out new ways of carrying out "ethical work" (Foucault 1990, 2000) in relation to the experiences and ontologies of the research participants that the ethnographer learns from. These two features clearly influenced the knowledge produced through my fieldwork and, specifically, the topic and focus of my dissertation, which formed the basis of this monograph. However, if in my writing I seem to privilege the positive aspects of the lives of my participants and their ways of finding empowerment in a variety of situations, this should not be regarded as either being solely due to my own optimism or as reflecting the unlikely condition that the people I met during my fieldwork had been spared from going through difficult life experiences. Rather, this is because they emphasized the full half of the glass in their relationships with me. I learnt from my participants to be even more optimistic, maybe a bit more laidback, to treat my domestic others with kindness, and to enjoy the nice weather while it's here.

Besides being an optimist, I am also an avid reader and amateur writer of fiction. As an aid in imagining how other people live and understanding their lives, fiction has always been a cherished resource for me. During my time in England, I serendipitously came across books—novels and autobiographical memoirs—that will always stay with me and whose company I cannot disentangle from the experience of doing my fieldwork and the process of writing this monograph. I mention the names of the writers with gratitude: William Cooper, Janice Galloway, Andrea Levy, Blake Morrison, Dodie Smith, Nina Stibbe, Jonathan Taylor, Sue Townsend, and Jeanette Winterson.

LIST OF FIGURES

CHAPTER 1

Introduction: The Time We Have–The Time We Make

A sense of urgency in relation to change is increasingly becoming part of everyday life for people all around the world who take part in political action or just witness it. This sense of urgency, which refers to both the seriousness of irreversible climatic transformations and to the necessity of immediate economic and social changes to stop the overproduction and overconsumption of goods and transport services globally, is what now connects the once distinct realms of nature, society, economy, and politics into one particular ontology. This shift suggests an emergent understanding of the fact that by radically transforming one element, such as the economy, it would be possible to influence the others. As the political commentator Naomi Klein (2014) argues, global warming can be regarded as a direct consequence of implementing and following, for over a century, a capitalist economic model focused on indefinite growth, which we now have the chance to change by opting for and developing alternative models, such as economic degrowth (D'Alisa et al. 2015). The consensus over the fact that climate change is actually happening introduces what Knox describes as a particular ontology of climate as systemic interconnection that establishes an imperative for people to acknowledge their agency, "albeit at the moment at which the possibility of that agency making a political difference seems about to disappear" (2015: 105). The urgency of change, therefore, articulates new considerations of human agency as well as of time.

© The Editor(s) (if applicable) and The Author(s) 2016
R. Moroşanu, *An Ethnography of Household Energy Demand in the UK*, DOI 10.1057/978-1-137-59341-2_1

In the last two decades, vitalist (Latour 2005; Massumi 2002; Thrift 2008) and posthumanist (Haraway 1991) approaches in social sciences—related to what Boyer (2014) calls the "anti-anthropocentric" turn in human sciences—aimed to de-centre the human subject, insisting instead on the agency of objects and materials and on the vitality of a world made of indeterminate relations that are autonomous of their subjects. These theoretical advances, however, came with a tendency to disregard the role of human creativity in making relationships and meanings, as Moore (2011) argues, while also dismissing any form of inquiry into human reason and agency "as though that inquiry were itself somehow part of the problem rather than a complementary project of truth-finding, or, better still, part of the solution to our contemporary challenges" (Boyer 2014: 319). However, calls for a reconceptualization of human agency have recently started to emerge (Gershon 2011; Moore 2011), some even from social theorists who were at the forefront of its denouncement (Latour 2014). It is only through human action that the politico-economic change Klein calls for could emerge. And one of the most effective ways for scholars in the social sciences to contribute to making this change possible is by creating a conceptual space where human agency is acknowledged while being defined in much more ways than one.

With the dismissal of human agency in the last two decades, time itself has lost an important dimension: the "near future" (Guyer 2007), a temporal frame that allows shaping—the idea that human beings can modify the future through their significant social and political engagements. We are left instead with only two temporal modalities, which Guyer calls fantasy futurism and enforced presentism. The latter can be regarded as emerging in relation to a narrative of acceleration characterized by "time-space compression" (Harvey 1990), instantaneous time and increased mobility (Urry 2000), and a network society of "timeless time" (Castells 1996) where new information technologies ask for simultaneity (Eriksen 2001) and where the markets of global finance function at very high speeds (Virilio 2012).[1] According to this frame, the present is fast and encompassing in a way that does not leave room for people to think about tomorrow or about what they would like to change in their lives or in the world. The increasing pace of everyday life acts as an anaesthetic for human agency: one's capacities for social critique are put to sleep in the face of a constant preoccupation with immediate survival. Fantasy futurism, then, emerges as a way of telling oneself that things will turn out fine, that the world will take care of itself: future technological innovation will solve all the problems related to the depletion of Earth's natural resources, and financial markets will surely find their way back to a stable course if left to their own devices.

The possibility of re-inhabiting and re-appropriating the near future is, therefore, related to finding ways for reconsidering and for reconceptualizing human agency in scholarship as well as in the world. Such an attempt is proposed here through the development of an analytical framework for looking at everyday actions of sustainability, as well as at everydayness more generally, by acknowledging their contribution to social change. The "ordinary agency" of everyday life—related to tacit, non-intentional, and unconscious acts of creating meaning such as imagination and taking action on impulse—will be described in correspondence with a set of temporal modalities enacted in domestic settings: spontaneity, anticipation, and "family time." Theoretical support from Foucault's (1990, 2000) work on ethical practices, as well as from recent anthropological re-readings and advancements of this work (Faubion 2011; Laidlaw 2014; Moore 2011), will be sought in an endeavour to propose an "ethical turn" in the study of sustainability and ethical consumption.

ENERGY, SUSTAINABILITY, AND SOCIAL SCIENCES

More than other forms of social inquiry on topics related to sustainability, contemporary energy research places a special emphasis on the relationship between energy, power, and politics. This is because the arena of global energy choices is currently being configured remotely from the experience of everyday life as well as from the opinions and beliefs of laypeople on this matter. Growing one's own food organically in a garden or allotment or cycling to work instead of driving are decisions that anybody could make, more or less, any time. However, allowing or banning fracking for shale gas near and under National Parks and other protected areas in the UK, for example, is a decision that is solely in the hands of the national government.[2]

Recent advancements in energy scholarship have examined various ways to theoretically articulate the actual relationships between energy infrastructures and institutions of political power. Following Foucault's concept of biopower, Boyer (2014) introduces "energopower" as a conceptual frame for an analytical method of understanding modern power. From a slightly different theoretical perspective, bringing together the entropy law of thermodynamics and Marxist and ecological economics, Hornborg (2013) proposes a radical reinterpretation of technology as the displacement of slavery by showing that, in substituting organic for inorganic energy, the industrial revolution created the illusion that the labour of oil and of other natural resources need not be remunerated at the same costs that the human labour necessary to accomplish a similar job would be. The rethinking of modern power in relation to energy infrastructures that Boyer and Hornborg, among others,[3] intelligently propose could lead to a realization of the fact that, at present, global energy choices need to be regarded as the ultimate terrain for political contestation and activism. In Stirling's (2014) view, there are two radically contrasting paths emerging from this terrain. The first path is a "progressive" transformation focused on renewable energy resources and technologies and on addressing wider sustainability benefits while eliminating carbon emissions. The second path follows a "business as usual" scenario, in which there is no effort to reduce carbon levels. This scenario proposes a "conservative" transformation focused on climate geoengineering interventions, such as carbon capture and storage, aimed "solely at assuming human 'control' over the planetary climate" (Stirling 2014: 85) and leaving energy needs unaddressed.

In this context, looking at the everydayness of energy and the ways in which laypeople use energy-consuming devices in their homes—which is what my research did—is an ambivalent endeavour. First, it is, of course, tremendously important to know more about the ways in which people understand, use, and value energy as a cultural artefact (Strauss et al. 2013) or as what Wilhite (2005) calls a quintessential social good. Secondly, there is an immense asymmetry of power between, for example, the situation of choosing an energy-saving light bulb over a traditional one and the situation of authorizing a tar sands oil-extraction project over a solar power plant.[4] During my fieldwork, the people who took part in my research, as well as myself, were aware of this asymmetry of power, and they brought it up in one-to-one encounters and in more public meetings, such as during a feedback event that marked the end of the project, which will be discussed in Chap. 4.

Sustainability, as a wider context for energy research, has been a topic of long and sustained interest in the social sciences (Milton 1996). One could argue that some of the main questions in this field, such as how to move towards a pro-environmental society, cannot be disentangled from the issue of responsibility: Who should be responsible for proposing, organizing, and producing this move? Hargreaves (2012) shows that early research on pro-environmental behaviour, following environmental psychology approaches, framed the problem as a result of faulty human decision making and thought the resolution resided solely in providing more information on the topic. Directed towards individual citizens, this approach was widely adopted in policymaking by providing environmental education through information campaigns with the purpose of changing behaviours. This top-down process of placing all responsibility on individuals has been criticized from the vantage points of other social sciences paradigms, such as theories of governmentality and biopolitics and social practice theory, for trying to impose a vision of the "green citizen" (Hobson 2013) as a "carbon-calculating individual" (Slocum 2004) while positioning governments "as enablers whose role is to induce people to make pro-environmental decisions for themselves and deter them from opting for other, less desired, courses of action" (Shove 2010a: 1280). Moreover, the influence of behaviour-change approaches on the overall sustainability research agenda and their relationships with policy-making, have been denounced as problematic because they support existing political interests by promoting a simplistic model of social change (Shove 2010b). To surpass the limitations of a behaviour change model while also not losing sight of a sought-after pro-environmental transformation, other forms of social theory have been called for. Social practice approaches to consumption, as some of the most vocal responses to the dominance of behaviour change paradigms in the UK, move the focus from the individual to the ways in which energy-consuming practices, such as cleaning (Shove 2003), come into being and are shaped and normalized in relation to specific socio-economic and technological conditions. Social change towards sustainability, therefore, is not a matter of changing individual behaviour but of changing the socio-material conditions—values as much as infrastructures—that frame practices. While behaviour change approaches regard people as "autonomous agents of choice and change" (Shove 2010a: 1279), social theories of practice regard people as carriers of practice.

The ethos conveyed by social practice theorists was inspiring for my work when I entered this debate. However, while the socio-material conditions of energy-consumption practices are slow to change, the question of responsibility still remains. Is it realistic to expect governments to make the best decisions and make energy infrastructures eco-friendly? Or might it be that environmental concerns are not the only or the main considerations that they follow during the decision-making processes? If urgent change is needed, then activism might be more influential to political decision-making than social theory is. For this, social theorists need to give people back their agency in the ways they describe the world. A renewed attention to agency does not limit this concept to a capacity for choosing between low-carbon and carbon-intensive practices and goods, as it would be in a behaviour change model; instead, human agency is what can call for and effect radical social transformation, as was the case with suffrage movements and the Civil Rights Movement.

The human geographer Kersty Hobson argues for developing a conceptual space in social theory in which everyday practices associated with domestic sustainable consumption are not "inherently devoid of a worthwhile form of personal and material environmental politics" (2011: 206). This would be a way of looking at people neither as the carbon-calculating citizens of behaviour change paradigms, nor as the sole carriers of practices of practice theory. She suggests that some possible theoretical tracks to follow in this endeavour would be inspired by Mouffe's (2005) conceptualization of a broader sense of the political, as well as by Foucault's later writings on ethics and self-formation.

Here, I follow Hobson's call for proposing a way of acknowledging the significance of everyday practices of sustainability by developing a theoretical framework inspired by Foucault's work on ethical practices. I will show that this is also a way of bringing agency back to human action and, therefore, of providing a resolution to the ambivalent endeavour of researching domestic energy demand. While remaining an important issue that needs to be pursued, the asymmetry of power between the everyday actions of laypeople and the power of governments to decide global energy choices stops providing onus to conducting research if everyday actions are regarded as expressive of human agency and as affording political impact. In following a Foucauldian apparatus for looking at ethical practices, my approach to practice is different than the way in which social practice theories conceptualize practices. This is because the theoretical framework developed by Foucault advanced Aristotle's conception of *praxis* and his definition of actions of "doing" (Faubion 2011; Laidlaw 2014), which differs considerably from Bourdieu's (1977, 1979) theory of practice that was, arguably, seminal for the development of social practice theoretical approaches to consumption (Warde 2005). This point, among others, will be addressed in Chap. 2.

THE CONTEXT OF MY RESEARCH

This monograph discusses my doctoral research, which was conducted as part of a wider interdisciplinary research project, namely, the Low Effort Energy Demand Reduction (LEEDR) project. Based at Loughborough University and funded by the Engineering and Physical Sciences Research Council (EPSRC) of the UK, LEEDR brought together fifteen researchers from five departments: building engineering, systems engineering, computer science, design, and social sciences. The project investigated the domestic energy consumption of UK households with the intent of proposing digital interventions that would help lower energy demand. From the start, the focus of the project was twofold: to reduce energy demand and understand the ways in which families used digital media devices so that interventions based upon these technologies could be developed. Both of these objectives, in conjunction with the general domestic sustainability research agenda related to existent concerns in national environmental policy, framed my research from the start and influenced the way in which I developed my research design.

One of the current research areas that the EPSRC allocates funding for is energy, which is related to the efforts for meeting the environmental targets and policy goals set by the Climate Change Act. Adopted in 2008, the Climate Change Act legally binds the UK's Parliament to reduce the country's carbon emissions by 80% in relation to the 1990 baseline by the year 2050. According to a set of energy consumption scenarios produced by the Department of Energy and Climate Change (DECC), it is estimated that a reduction of 26–43% in energy consumption will be required to achieve the 80% reduction in carbon emissions. While accounting for approximately 30% of the UK's greenhouse gas emissions, domestic energy consumption at the moment is not lower than in 1990. The predictions of the Energy Saving Trust for electricity consumption show that "domestic electricity demand in 2020 is forecast at 80 TWh, a reduction in real terms of about 6% from 2009 levels, but 14% higher than 1990" (EST 2011: 29). This context explains the interest in funding research in the area of energy efficiency, which was addressed by the LEEDR project.

Working as part of an interdisciplinary project involved participating in fortnightly meetings and thematic day workshops with the entire LEEDR team to develop interdisciplinary dialogues. I collaborated with the other team members in designing and conducting the research, discussing and analysing the data, and writing journal and conference papers. At the same time, I developed my doctoral project independently, introducing additional research methods and drawing upon additional bodies of literature from the fields of the anthropology of time, applied anthropology, and the anthropology of Britain. While maintaining interest in the digital media usages expressed initially by the LEEDR project, I approached these questions in relation to the temporal dimension of domesticity. Moreover, after finalizing my ethnographic fieldwork and after analysing my findings within a social anthropological framework, I returned to the concern of reducing domestic energy consumption embedded in the LEEDR project. I made an exercise of formulating possible applications for reducing energy demand that could result from the knowledge produced through my fieldwork. The ideas that emerged from this exercise are discussed in the last chapter of the book.

FAMILY PARTICIPANTS

The LEEDR group conducted research on twenty self-selected family-participants who volunteered to take part in the project for two and a half years. Their participation was not incentivized and their reasons for volunteering were individual and diverse, such as an interest in living more sustainably or lowering their energy bills. They agreed to have their energy consumption monitored for the entire duration of their participation and to take part in ethnographic research and research conducted by colleagues from the discipline of design. The participant families were all home-owners; eighteen families could be described in terms of income and education as "middle-class," whereas the other two families could be described as "working-class." With the exception of one single-parent family and one extended family where the maternal grandmother was live-in, all the family-participants were two-parent nuclear families. Moreover, all adult couples were heterosexual and married. Each family had between one and four children with ages ranging from one to twenty two.

While the LEEDR recruitment materials did not contain any descriptive cues regarding the types of families wanted, except for the essential criteria of being home-owners so that it was possible for them so that it was possible for us to install monitoring equipment, it turned out that all the families who volunteered reflected what can be called a traditional family ideology, namely, a heterosexual married couple with children. James (1998) argues that the rhetoric of the traditional family in the UK continues to be conferred with ideological solidity by social policy and political practice "despite its considerable inappropriateness for accounting for the contemporary forms of family life" (James 1998: 143). The dominance of this rhetoric might be one of the reasons why other alternative forms of families were not represented in this research, as the apparently neutral descriptive cue of "home-owner family" might have automatically triggered the image of a "cereal box" middle-class family, which many potential participants might not have identified with.

The fact that all the families had children relates with existing cultural understandings of kinship as well. In her study of childhood identity in Britain, James discusses the assumption that "parenting is what transforms the couple into a family—a shift that is also popularly held to make a house into a home" (1998: 144). In other words, UK adults need to have children to be regarded as having a "family" or a family-style lifestyle (Strathern 1992). This social requirement is associated with a set of moral values that, in a secular context, has the potential to produce a dominant and self-sufficient model of morality. As Strathern (1992) argues, in post-Thatcherite Britain, by "doing" convention, which here is opting for a family-style lifestyle, the individual "shows his or her capacity for morality, and thus makes explicit the fact that moral behaviour is contingent on the capacity for choice" (1992: 162). In Chap. 7, I discuss the complex relationships that my participants had with the ideologically solid rhetoric of the family and that they expressed in "family practices" (Morgan 1996) of making "family time" that both showed their "doing" of convention and their creativity in challenging it. Further, in the last chapter, I draw attention to a potential conflict between the morality associated to having a family-style lifestyle and the morality of sustainable and low-carbon living practices.

An Interest with Time

A focus on time emerged from the early stages of my research, from pilot interviews conducted as part of the LEEDR project, and from my diary that recorded my initial surprise and observations related to just having had moved from Bucharest, Romania, to a small town in the UK.

Later, in my fieldwork with English middle-class families, time appeared as one of the main constraints on people's lives. It was one of the main agents that stood in the way of them achieving a sense of having the life they always wanted. The state, political power, the economic system, financial constraints, gender inequalities, and specific sets of social norms and expectations were not apparent sources of conflict. Time, however, was. Lack of time was what all the people taking part in my research felt stopped them from being what they aspired to be—from pursuing all the hobbies that they were interested in, from making their house into an ideal home, or from modelling their children in a way that would fully represent their values and beliefs. This sense of a lack of time emerged most strongly when I asked the research participants if there were ever moments when they wanted to make time go faster or when they felt bored. The unanimous response to this question was a strong no, accompanied by surprised laughter and various explanations, such as Vic's, below[5]:

> Normally, time is something that we don't have enough of anyway. So, to make it pass quicker, it's not something we would, generally, want to do. I can't remember the last time I was bored. I can remember the last time I was thinking "why am I watching this?" but not actually bored, no. We're always busy; if we're waiting twenty minutes for something to be ready in the oven, then we would use that twenty minutes for something else. (interview Vic and Gail)

For the people taking part in my research, time was running fast, which was visible in the changes in their children as they grew up; time was finite, an observation which usually emerged on Sunday evenings and at the end of holidays. Time was not what it used to be. When I asked Elaine and Chris if they had a routine of using music, radio, or the TV as a background to set the atmosphere for specific collective family activities, they said that they did not follow a routine as their parents might have done by listening to the same radio show or by watching the same TV soap opera every evening. They were not trying to create a sense of continuity and repetition in the ways they articulated domesticity, but to just get done all the evening tasks that they needed to complete for themselves and for their children, as Chris explains:

> I guess the modern world and family processes, they don't really lend themselves to the kind of relaxed, context repetition of previous generations. I mean, we're just doing whatever we need to do to get everything done at the moment, aren't we? (interview Chris and Elaine)

For the majority of my research participants, everyday domestic life was far from routinized; it was rather made of a series of spontaneous decisions. These decisions generally followed a scenario: What's the situation, what's the time, what needs to be finished, what can be paused, and what can be left for tomorrow? The increased busyness of life as well as the blurring of the boundaries between work and home facilitated by the new affordances of digital technologies—for example, instantly providing access to work emails on smartphones and tablets—made people constantly feel limited by the amount of time available. Multi-tasking and switching between activities while waiting were two techniques for gaining time. Here, Chris describes his strategies for not wasting precious time:

> You can never wait for anything; you start something off and then you think, "Right, I'll go and do this." And, then, you start something else off, and then you go and do something else and start that off, and then, all of a sudden, you're going, "Oh, there was something else I needed to do and that's going to be ready in a minute and I've got to get the ..." I mean, I've even got to the point of trying to write down what comes into my head that I should be doing so that when I come back I won't forget what it was I thought of five items ago that I can do. I mean, the bizarre thing is, it's like putting your cup of tea in the microwave for thirty seconds or something—even that you can't stand to stand there and watch. You cannot watch for thirty seconds, you have to go and do something else ... I mean, watching seconds count down on the microwave is like watching your life disappear. They should put a skull on the front or something. (interview Chris and Elaine)

While this sense of passing through everyday life against the constant ticking of a timer was clearly related to the sheer busyness of life in modern times, I argue that one important element in generating and sustaining this feeling was an essential change in the visual representation of time measurement. In my fieldwork, the preferred method for telling the time during the day was by checking one's mobile phone or computer. These devices display time as a series of four numbers—for example, 15:52—the last of which changes every minute. Analogue clocks represent time as cyclical: the speed of the moving hands is constant and their movement visible, which makes it possible for one to see the choreography of the passing of one minute, to anticipate a future event that will start after a specific number of set segments of time have passed, and to visually "calculate" the amount of time that is left before one would need to leave for work. In a digital numeric representation, time appears as linear and as moving forward like a chronometer rather than in circles and clockwise; this representation appears as a given and unquestionable set of numbers that do not permit a creative interpretation, anticipation, or waiting, as Chris mentions. The way in which digital clocks show time is similar to how time's passing is displayed on a time bomb in an action movie. No wonder that one cannot stop the chain of activities in order to just wait for something to be ready. Digital representations of time are increasingly replacing analogue ones by being embedded in various domestic appliances, from alarm clocks to TV boxes. This would mean that from the moment they wake up and until they go to bed, people situate themselves in relation to time by using a digital numerical framework of linearity, or progression.

When asked about the differences between using a digital and an analogue clock, some research participants said that it did not make any difference for them, because they were using both devices for an informative purpose, while others said they preferred to read time on an analogue clock face; however, due to the pervasiveness of digital devices, they did not use an analogue clock as often as they would have liked to. Digital clocks were associated with precision, and they were often checked because people knew that the time on mobile phones and computers was synchronized through satellites, showing the "right" time that was used globally. Analogue clocks, on the other hand, were prone to error. Their invention is related to a previous scientific era when the time on mechanical clocks was calibrated to the cycles of caesium atoms and not to the rotation of the Earth, as discussed by Birth (2012). Steve told me that he prefers to read time on an analogue clock face rather than by using a digital clock:

> Because, at a glance, you know what you're looking at … When I see the hand, for me, it equates to an amount of time. Certainly, in the morning, if I need to take Alan at eight o'clock to school, and I'm sitting here, and I've got fifteen minutes left to get dressed and get out. I do see that more relevant as a sequence of time, than if I've looked and it said "45." Sometimes I get caught out a little bit, if I've used the clock on my car for instance: what time you've got left to get somewhere. Sometimes I've got caught out, 'cause somehow it doesn't… I don't know, it goes faster; for me, anyway. (interview Steve and Iris)

As he is not able to see the time passing on the digital clock in his car in the way that he does when using an analogue clock, Steve feels that time goes faster. His driving is not attuned to a visible movement of cyclical time, rather it is parallel to and in competition with a progressive linear invisible time that is expressed through a set of numbers that can be trusted to represent "reality" and that appear as an unquestionable verdict.

The cognitive anthropologist Kevin Birth argues that the time-reckoning tools developed by humans can be regarded as cognitive artefacts, objects that people use to think about the world. Clocks and calendars, which provide a set of dominant representations of time, are tools that humans have placed knowledge in and that they have come to rely on instead of continuing to develop and use their cognitive skills for telling the time. "These skills can be cultivated, but normally are not, because their cultivation creates a sense of multiple temporalities of which clocks and calendars represent only two. To place these alternate temporalities in counterpoint with the dominant modes of reckoning time challenges these dominant modes" (Birth 2012: 11). Following Birth's approach, one can argue that digital clocks are cognitive artefacts people use to think about the world and that influence the ways in which they position themselves in relation to time. Digital clocks are not accountable for what is regarded as the increased pace of everyday life in modern times, but their pervasiveness can be considered to contribute to maintaining one's perception of competing against the counting down of a timer as a way of living. While analogue clocks, which are prone to error and require resetting periodically, leave room for the imagination that time could pass at different speeds and sometimes even stop, digital clocks, which are synchronized to satellites, are never wrong. In fact, digital clocks privilege a specific type of engagement, one based on linear progression and not on imagination or daydreaming.

However, I argue that the constant use of digital clocks as dominant time-reckoning tools does not completely erase alternative human skills for telling time, for imaging time, and for engaging with time. Rather than being absent, these skills are exercised tentatively and informally in the domain of everyday domestic life, as well as in other domains.

Anthropologist Carol Greenhouse (1996) argues that temporal modalities are always linked to understandings of agency, because human action is deemed relevant or not in relation to time. By following this idea, three modalities of time will be described and conceptualized in three dedicated chapters: spontaneity, anticipation, and family time. I argue that these time modes express counter-discourses to the dominance of linear time—which is articulated and maintained through digital clocks, among other techniques—while also suggesting specific understandings of agency that are enacted in everyday life. Part of the discussion will be focused upon examples where these alternative time modes are articulated through the use of digital media, showing that people have the capacity and the creativity to employ the very tools that embody a linear representation of time to create counter-discourses to this form of temporality. These counter-discourses, however, should not be regarded as forms of resistance to linear time but as co-existing with it. By employing alternative temporalities, people momentarily ignore the requests and the limitations associated with a linear temporal framework. Instead, they express alternative ways of engagement with the world articulated through "ethical imagination" (Moore 2011) and through the development of new forms of sociality.

THE CHAPTERS

Chapter 2 discusses the overall theoretical framework of the book. Drawing on the calls by Moore (2011) and Hobson (2011) for a new recognition of the potentialities of human agency, the chapter proposes the concept of "ordinary agency" (1) in relation to recent advancements of Foucault's work on ethical practices in the field of the anthropology of ethics (Faubion 2011; Laidlaw 2014), (2) as a response to Gershon's (2011) critique of "neoliberal agency," and (3) in relation to understandings of time as discussed by Greenhouse (1996).

Chapter 3 introduces the ethnographic location where the research was conducted: a provincial medium-sized town situated in the Midlands area of England. It then situates the field site in relation to conceptualizations of locality and place in the anthropology of Britain. It continues by discussing the applied interdisciplinary context of the research, and it introduces a set of participant-led and arts-based methods that were employed in this research as part of long-term ethnographic fieldwork.

Chapter 4 acquaints the reader with the Smiths, the Nicholls, the Loves, the Hewitt-Mitchells, and the Johnsons. The ethnographic discussion is developed through the experiences of these five families in the three following chapters.

Chapter 5 addresses the concept of spontaneity as a mode of action and as an approach to time. A discussion of the ways in which achieving immediacy can contribute to extending one's ethical imagination is developed, and an analytical framework for spontaneity is proposed by drawing upon the work of the British philosopher Elizabeth Anscombe on intentionality (1957).

In Chap. 6, the temporal modality of anticipation is introduced in relation to the concept of Mother-Multiple. The techniques of imagination employed in short-term anticipation and in "looking forward to" anticipation, as well as the effects on energy consumption that these approaches to time entail, are considered. The Mother-Multiple concept is developed to explore enactments of care oriented towards one's "domestic others": family members, pets, or the home itself as an entity.

Chapter 7 discusses the time mode of "family time" together with two specific forms of domestic sociality that are both enacted through the use of digital media. In relation to English middle-class kinship (Strathern 1992), family time emerges as a concept that people use to "measure" the quality of home life.

Besides developing conceptualizations for regarding three specific temporal modalities and the forms of ordinary agency they engender, these chapters also focus upon three different scales of analysis. Chapter 5 looks at individuals, Chap. 6 addresses relationships between two parts—the Mother-Multiple and the "domestic others"—while Chap. 7 looks at commonality as well as specific forms of sociality of togetherness that are articulated in family settings.

Chapter 8 focuses on potential applications for the anthropological knowledge that was co-produced in the ethnographic fieldwork and in the analytical process. Two sets of propositions for changing the types of questions related to the domestic sustainability research agenda are developed.

Notes

1. However, approaches that insist one-dimensionally on acceleration have been criticized as being technologically deterministic and lacking empirical evidence (Glennie and Thrift 2009); such evidence would point instead towards the diversity of contemporary experiences of time (Wajcman 2015) and a multitemporality approach (Serres 1982).
2. After allowing fracking near and under National Parks through horizontal drilling (Carrington 2015), the UK government is now considering allowing more forms of fracking in those areas (BBC 2015).
3. In rethinking the history of energy, Mitchell (2011) develops a powerful and thought-provoking account of the entanglements between modern democratic power and carbon energy systems.
4. One is allowed to suspect that the letter decision might be dictated by commercial and political interests rather than by scientific calculations and mod-

els, as there is already important evidence that by using currently available technological innovations it would be possible to power the planet exclusively from renewable sources by the year 2050 (WWF 2011; EREC 2010; PWC 2010).
5. All names are pseudonyms.

REFERENCES

Anscombe, Gertrude Elizabeth. 2000 [1957]. *Intention.* Cambridge, MA: Harvard University Press.
BBC. 2015. "Government Considers Fracking Plans in National Parks." http://www.bbc.co.uk/news/uk-34718984
Birth, Kevin K. 2012. *Objects of Time: How Things Shape Temporality.* New York: Palgrave Macmillan.
Bourdieu, Pierre. 1977. *Outline of a Theory of Practice.* Cambridge: Cambridge University Press.
Bourdieu, Pierre. 1979. *Distinction: A Social Critique of the Judgement of Taste.* London: Routledge.
Boyer, Dominic. 2014. "Energopower: An Introduction." *Anthropological Quarterly* 87 (2): 309–333.
Carrington, Damian. 2015. "Fracking Will Be Allowed under National Parks, UK Decides." *The Guardian*, February 12. http://www.theguardian.com/environment/2015/feb/12/fracking-will-be-allowed-under-national-parks
Castells, Manuel. 1996. *The Rise of the Network Society.* Oxford: Blackwell.
D'Alisa, Giacomo, Federico Demaria, and Giorgos Kallis. 2015. *Degrowth: A Vocabulary for a New Era.* New York: Routledge.
EREC. 2010. *Rethinking 2050: A 100% Renewable Energy Vision for the EU.* Brussels: EREC.
Eriksen, Thomas. 2001. *Tyranny of the Moment: Fast and Slow Time in the Information Age.* London: Pluto Press.
EST. 2011. *The Elephant in the Living-Room: How Our Appliances and Gadgets Are Trampling the Green Dream.* http://www.energysavingtrust.org.uk/Publications2/Corporate/Research-and-insights/The-elephant-in-the-living-room
Faubion, James D. 2011. *An Anthropology of Ethics.* Cambridge: Cambridge University Press.
Foucault, Michel. 1990. *The History of Sexuality*, vol. 2. London: Penguin Books.
Foucault, Michel. 2000. *Ethics: Subjectiviy and Truth (Essential Works of Foucault 1954–1984).* London: Penguin Books.
Gershon, Ilana. 2011. "Neoliberal Agency." *Current Anthropology* 52 (4): 537–555.
Glennie, Paul and Nigel Thrift. 2009. *Shaping the Day: A History of Timekeeping in England and Wales 1300–1800.* Oxford: Oxford University Press.

Greenhouse, Carol J. 1996. *A Moment's Notice: Time Politics Across Cultures.* New York: Cornell University Press.

Guyer, Jane I. 2007. "Prophecy and the Near Future: Thoughts on Macroeconomic, Evangelical, and Punctuated Time." *American Ethnologist* 34 (3): 409–421.

Haraway, Donna. 1991. "A Cyborg Manifesto: Science, Technology, and Socialist-Feminism in the Late Twentieth Centurt." In *Simians, Cyborgs, and Women,* edited by Donna Haraway, 149–181. London: Free Association Books.

Hargreaves, Tom. 2012. "Questioning the Virtues of Pro-Environmental Behaviour Research: Towards a Phronetic Approach." *Geoforum* 43 (2). Elsevier Ltd: 315–324.

Harvey, David. 1990. *The Condition of Postmodernity.* Oxford: Blackwell.

Hobson, Kersty. 2011. "Environmental Politics, Green Governmentality and the Possibility of a 'Creative Grammar' for Domestic Sustainable Consumption." In *Material Geographies of Household Sustainability,* edited by Ruth Lane and Andrew Gorman-Murray, 193–210. Farnham, UK: Ashgate.

Hobson, Kersty. 2013. "On the Making of the Environmental Citizen." *Environmental Politics* 22 (1): 56–72.

Hornborg, Alf. 2013. "The Fossil Interlude: Euro-American Power and the Return of the Physiocrats." In *Cultures of Energy: Power, Practices, Technologies,* edited by Sarah Strauss, Stephanie Rupp, and Thomas Love, 41–59. Walnut Creek, CA: Left Coast Press.

James, Alison. 1998. "Imagining Children 'At Home', 'In the Family' and 'At School': Movement Between the Spatial and Temporal Markers of Childhood Identity in Britain." In *Migrants of Identity: Perceptions of "Home" in a World of Movement,* edited by Nigel Rapport and Andrew Dawson. Oxford: Berg.

Klein, Naomi. 2014. *This Changes Everything: Capitalism vs. the Climate.* New York: Simon & Schuster.

Knox, Hannah. 2015. "Thinking like a Climate." *Distinktion: Scandinavian Journal of Social Theory* 16 (1): 91–109.

Laidlaw, James. 2014. *The Subject of Virtue: An Anthropology of Ethics and Freedom.* Cambridge: Cambridge University Press.

Latour, Bruno. 2005. *Reassembling the Social: An Introduction to Actor-Network Theory.* Oxford: Oxford University Press.

Latour, Bruno. 2014. "Agency at the Time of the Anthropocene." *New Literary History* 45 (1): 1–18.

Massumi, Brian. 2002. *Parables for the Virtual: Movement, Affect, Sensation.* Durham, NC: Duke University Press.

Milton, Kay. 1996. *Environmentalism and Cultural Theory: Exploring the Role of Anthropology in Environmental Discourse.* London: Routledge.

Mitchell, Timothy. 2011. *Carbon Democracy: Political Power in the Age of Oil.* London: Verso.

Moore, Henrietta L. 2011. *Still Life: Hopes, Desires and Satisfactions*. Cambridge: Polity Press.

Morgan, David H.J. 1996. *Family Connections: An Introduction to Family Studies*. Cambridge: Polity Press.

Mouffe, Chantal. 2005. *On the Political*. New York: Routledge.

PWC. 2010. *100% Renewable Electricity: A Roadmap to 2050 for Europe and North Africa*. London: PWC.

Serres, Michel. 1982. *Hermes: Literature, Science, Philosophy*. Baltimore: John Hopkins University Press.

Shove, Elizabeth. 2003. *Comfort, Cleanliness and Convenience: The Social Organization of Normality*. London: Berg.

Shove, Elizabeth. 2010a. "Beyond the ABC: Climate Change Policy and Theories of Social Change." *Environment and Planning A* 42 (6): 1273–1285.

Shove, Elizabeth. 2010b. "Social Theory and Climate Change: Questions Often, Sometimes and Not yet Asked." *Theory, Culture & Society* 27 (2–3). TCS Centre: 277–288.

Slocum, Rachel. 2004. "Consumer Citizens and the Cities for Climate Protection Campaign." *Environment and Planning A* 36 (5): 763–782.

Stirling, Andy. 2014. "Transforming Power: Social Science and the Politics of Energy Choices." *Energy Research and Social Science* 1. Elsevier Ltd.: 83–95.

Strathern, Marilyn. 1992. *After Nature: English Kinship in the Late Twentieth Century*. Cambridge: Cambridge University Press.

Strauss, Sarah, Stephanie Rupp, and Thomas Love. 2013. "Introduction. Powerlines: Cultures of Energy in the Twenty-First Century." In *Cultures of Energy: Power, Practices, Technologies*, edited by Sarah Strauss, Stephanie Rupp, and Thomas Love. Walnut Creek, CA: Left Coast Press.

Thrift, Nigel. 2008. *Non-Representational Theory: Space, Politics, Affect*. London: Routledge.

Urry, John. 2000. *Sociology Beyond Societies: Mobilities for the Twenty-First Century*. London: Routledge.

Virilio, Paul. 2012. *The Great Accelerator*. Cambridge: Polity.

Wajcman, Judy. 2015. *Pressed for Time: The Acceleration of Life in Digital Capitalism*. Chicago: University of Chicago Press.

Warde, Alan. 2005. "Consumption and Theories of Practice." *Journal of Consumer Culture* 5 (2): 131–153.

Wilhite, Harold. 2005. "Why Energy Needs Anthropology." *Anthropology Today* 21 (3): 1–2.

WWF. 2011. *The Energy Report: 100% Renewable Energy by 2050*. Gland: WWF.

CHAPTER 2

How the Light Gets In: A Theoretical Framework for "Ordinary Agency"

We did it in stages. Because we first moved in and, before any planning, we knocked these double doors through. That was the first thing we did. And we immediately saw light coming in into that room that had no light. If you imagine, in the hall it was a room: there was a ceiling and no windows, no daylight. And the only daylight in that room came from the front door. So then we knocked this through, and that just flooded it with light. And then we knocked another door through, and that flooded it with light. And then, the more we did it, gradually, we thought: "Oh, yeah, this is great! Yeah, this is lovely; this will be good". So yeah, we kind of built up. We didn't really know. It sort of evolved. (interview Iris and Steve)

While telling the story of their house extension, Iris arose from her chair. She pointed towards the hall with large gestures that helped her explain the complicated transformations their house went through after they bought it. We are all gathered around the kitchen table: Iris, her husband, Steve, me, and a fellow researcher. We discussed Iris's and my previous encounter, when we took a "house video tour" (Pink 2004) and explored their home and the ways in which Iris engaged with it.

After Iris, Steve, and their two children May and Alan, who were 11 and 13 at that time, had the chance to watch the nearly two-hour-long video recording, I visited them again for a follow-up interview, this time shadowed by a colleague. Iris's reflection on the process of transforming the ground floor came as a response to my insistent wish to find out how

© The Editor(s) (if applicable) and The Author(s) 2016   19
R. Moroşanu, *An Ethnography of Household Energy Demand in the UK*, DOI 10.1057/978-1-137-59341-2_2

it felt to completely change the whole layout of the house by moving the staircase, knocking down several walls, moving the main entrance door to a different side of the house, extending the kitchen, and practically changing all the ground floor rooms together with two bedrooms. Most of the families that I met during my fifteen months of fieldwork had stories of house transformations that followed the moment of acquisition, and they enjoyed recalling tearing down the wallpaper and redecorating in an attempt to get rid of any traces of previous ownership and make the house "their own."[1] Strathern (1992) interprets the belief that the interior of the house can be the object of personal, careful design as being in close connection to the modern English middle-class idea that the interior of an individual person is "an explicit object of improvement" (Strathern 1992: 102). The relationship between the interior of a person and the interior of a home has been inverted, as an interest in the good organization of the latter arrived to show the virtues of the former, for the new middle-class of the nineteenth century; therefore, "[t]he internal (what is within persons) has been literalised as an interior (residential) space" (Strathern 1992: 103).

In the case of Iris and Steve, the house transformation was substantial, and I wanted to know whether they could envisage it in advance. My questions made Iris bring the house plans to the table. They had someone who made the technical drawings, but they employed no architect. It was all based on their imagination and hope that the house could be transformed into their ideal home: a place where they could gather their extended family, which includes 35 people, for large parties and celebrations. They followed Iris's brother's advice to move the staircase, a transformation that he previously undertook for his own house and the results of which pleased him. The process of thinking, discussing, planning, drawing, and imaging the house took them approximately six months from the moment they moved in, and the building work took an additional six months. While showing me the floor plans and recalling this experience that happened five years before, both Iris and Steve admitted that they could not really envisage how the house would look like after the structural transformation by only looking at the drawings at the moment of planning. They just went for it, took the risk, and started knocking down walls one by one to let the light in.

Printed in large font on an A4 sheet of paper, Iris's words that opened this chapter have stood on the screen by my desk for over two years between photos of my friends and postcards. Leaving their place after the follow-up interview, I thought about the way Iris talked when she described the process of knocking down walls in order to let the light come in: she spontaneously stood up so as to embody the memory better. I transcribed that fragment the next morning. Her words felt powerful and inspiring, even if I did not know at that time how they were inspiring or whether and how I would later use them in writing about my fieldwork. These words are about an action of building work that one might imagine as dusty, noisy, and not generally very attractive to watch from close distance. What makes this event memorable? What makes the reminiscence of the light flooding the dark room memorable and shareable with a researcher?

In her recent publication on hopes, desires, satisfactions, and "the aspirational character of our relations to others, to knowledge and the world" (Moore 2011: 10), Henrietta Moore provides a critical review of the new wave of "vitalist" theories: theories of affect (Massumi 2002), actor-network theory (Latour 2005), and non-representational theory (Thrift 2008). Moore acknowledges the ontological break that vitalist theories have marked in relation to and as a critical response to social constructivist approaches by proposing "a decentring of the human subject, a recognition that the subject is formed through a series of encounters which it does not author or control" (2011: 180). These approaches move away from the human subject and outline the vitality of a world made up of indeterminate relations that are autonomous of their subjects (Massumi 2002; Thrift 2008). If one would follow the principles proposed by this new theoretical corpus, then it would be the element of light in Iris's account that would be emphasized as an agentic substance that autonomously floods the dark room; the intrinsic quality of immateriality that light possesses makes it possible for this substance to circulate, to sneak in. However, is it the special qualities of light that make the moment of flooding the dark room exceptional and memorable, or might it be the empowering freedom of knocking down walls that Iris wants to share with me?

During the follow-up interview, Iris and Steve recounted the way they spontaneously transformed the big open-space kitchen—their favourite room—into a "Christmas hub" during the last festive season. They connected the TV screen to the Internet through their video game console, and they accessed a video-sharing website to play a video of "a real crackling fire" while broadcasting a Christmas jukebox radio station by attaching speakers to their tablet. They laughed while recalling the episode, also feeling content with their creative invention. They could have gone into the living room and lit the real fire place, which they often use on winter evenings, but they created something different instead. The thorough details of their remembrance of this event suggest that the temporary technological assemblage was a product of their own making, just as knocking down the walls was, and embodied and expressed their subjectivities and agencies.

In concluding her book, Moore outlines that "perhaps it is too early yet to abandon a notion of the human subject marked by what is specifically human, most especially our desires, hopes, and satisfactions" (Moore 2011: 204). What she finds to be some important limitations of vitalist theories are an insistence on the autonomy of affect that dismisses the human creativity in making relationships and meanings and a disregard of the historical and cultural specificities of the contexts in which life takes place. Moore suggests that a new acknowledgment of human agency would be a possible way to surpass the dichotomous tensions between the social constructivist tradition and the new vitalist approaches. This shift would allow us to "think again" in a manner that recognizes human creativity and "ethical imagination." Moore's post-vitalist approach to human agency suggests that acts of ethical imagination of self-other relations or the question of "how we deal with each other" (2011: 15) are generative of social transformation. The concept of ethical imagination, which illustrates the forms and means "through which individuals imagine relationships to themselves and to others" (2011: 16), is proposed by Moore as an analytical tool to look not only at how actual relationships are experienced, but also at the new forms and potentialities for relationality that emerge from various encounters, such as those between human beings and new technologies.

We can now arrive to see Iris and Steve as agents who let the light come in by knocking walls down. Similarly, every morning, people—in Middleborough, in the UK, in the world—open their curtains and let the light in. It is this form of everyday human action that brings daylight into homes. People choose when to open and when to close curtains and blinds; their actions do not "produce" the light, but, more importantly, make manifest an engagement with the world that recognizes, accepts, and lets the light in.

These everyday actions make "morning" and "evening" happen inside people's houses. Subjective mornings and evenings can take place any time before or after their "natural" counterparts—but always in relation to them—and the times of their occurrences often vary with the day of the week. Several research participants mentioned that they used to close the curtains earlier on Saturdays than on midweek evenings, even though the sun did not set sooner on the weekends. Similarly, sometimes entire Sundays would be spent with the curtains closed in a 24 hour "non-day": a suspended and "stolen" time, pulled away from the ordinary rhythm imposed by clock and calendar time. This suggests that even if they might be analytically approached as routines, opening and closing curtains are not performed automatically without much thought at exactly the same times every morning and evening. They are actions that involve making decisions. These decisions are enacted by agents when they feel that it is the right time. The ways in which they make these daily decisions involve bringing together what is happening outside (getting brighter or darker) with what is happening inside (e.g., people still wearing their pyjamas in the morning or planning to "start the evening" sooner). Sometimes the actions of opening and closing curtains can be directed towards changing what is happening indoors or outdoors; for example, one can open the curtains to convince one's children that they need to get dressed or to make the daylight last longer during the winter.

To set her home in the mood for embracing the new day that has just started, it is important to Joyce to open the curtains on every single window in the house. However, one of the windows on the landing is particularly hard to reach because it is situated on the other side of the staircase. Joyce needs to lean over the baluster in order to reach the handle and open the blinds. Still, she performs this risky leaning twice every day: once for making morning and once for making evening. This action can be regarded as a brief form of worship that Joyce performs to assess through her body that a new day has come or is ending, respectively. Veena Das observes that "our theoretical impulse is often to think of agency in terms of escaping the ordinary rather than as a descent into it" (2007: 7). In Joyce's case, opening and closing down curtains are ordinary actions through which she takes and expresses control over her life by choosing when to make morning and evening happen.

The answer that I propose here to the question of how the light gets in, therefore, is that people let it in. Whether through opening curtains or through knocking down internal walls, people welcome and acknowledge the light if and when they wish to do so.

FOUCAULT'S ETHICAL PRACTICES

In her conceptualization of ethical imagination as an analytic tool, Moore (2011) draws upon Foucault's ideas about the relationships between ethical practices and the constitution of the knowing subject. She argues that by using the analytic lever of ethical imagination, one can arrive at new ways of looking at social transformation. The ways in which we deal with each other—or, in Foucault's original wording, "the strategies that individuals in their freedom can use in dealing with each other" (2000: 300)—are generative of social change, because the question of self-other relations stands at the very basis of what constitutes "the social." Moore's approach, therefore, surpasses a perspective on agency as being in opposition to structure, because the question of how we deal with each other can be seen as constitutive and as influencing both the direction and focus of individual actions and the ways in which we create social institutions, systems, and structures. She suggests that ethical imagination can be regarded as a primary site of cultural invention that is "brought into play by the advent of new information and new ideas, new ways of being and acting, new forms of representation and their mediation" (Moore 2011: 16) that, however, are not just linguistic or conscious.

Foucault's work on ethics develops the idea that ethical analysis is a mode of self-formation and a practice of freedom: "the freedom of the subject and its relationships to others ... constitutes the very stuff [*matière*] of ethics" (Foucault 2000: 300). In this approach, the relationship between ethics and freedom involves self-reflexivity, because "ethics is the considered form that freedom takes when it is informed by reflection" (Foucault 2000: 284). As a practice of freedom, ethics is directed towards limiting and controlling relationships of power that Foucault argues are widespread in all the forms of human relationships when one tries to control the actions of the other(s). James Laidlaw remarks that Foucault's merit is in rethinking the concepts of power and freedom in tandem rather than in opposition—freedom not as liberation from all social constraints but as an aspect of the way in which power relations are configured—and he argues that this rethinking could be an important resource for making freedom "both object and instrument of anthropological thought" (Laidlaw 2014: 92).

What Foucault calls practices of freedom should not, therefore, be seen as merely synonymous with what can be called resistance, even if these practices are a specific set of tools that human beings use for preventing an abuse of power. These are tools for processes of reflective self-formation, which are not only specific ways to prevent abuses of power, but primarily ways to improve "dealing with each other." The ethical for Foucault is thus the domain in which one can exercise a free relationship to the self (or *rapport à soi*), and the scope of ethical practice is to develop one's relationship to the self and to others.

Ethical practices are ethical in the sense that they represent the work that actors do in order to "make themselves into subjects of esteemed qualities or kinds" (Faubion 2011: 3). They are practical in an Aristotelian sense. In his approach to practice, Faubion (2001) argues, Foucault extends the Aristotelian distinction between doing and making. Aristotle draws this distinction by suggesting that "doing and making are generically different, since making aims at an end distinct from the act of making, whereas in doing the end cannot be other than the act itself: doing well is in itself the end" (Aristotle 1934: 337). For Aristotle, making is concerned with the creation of a new, finite object, while doing contains its scope in itself. Foucault's ethical practices are actions of doing that contain their scope in themselves. The scope is to engage in activities of self-reflection and not to craft out a new self, even if one inherently changes oneself and the ways one deals with the others through reflective actions.

Because Foucault's approach to practice gives a privileged place to change and, importantly, it provides "analytical space for attention to the subject's capacity to change itself" (Faubion 2011: 46), this represents a very different analytical project that the one undertook by Bourdieu in his conceptualisation of social practice. Bourdieu's approach, which led to the development of the paradigm of theories of social practice (Ortner 1984), proposes the concept of habitus that initially appears to be a development of Aristotelian practical reason as "the primary locus of structural reproduction" (Faubion 2011: 45). Bourdieu is concerned with the goals of practices: the social capital that can be derived, accumulated, and translated from engaging in various social practices (Lambek 2008: 136). These types of practices oriented towards ends are actions of making rather than of doing. Bourdieu's approach to practice, therefore, does not follow Aristotle's; rather, it can be argued it falls back into instrumentalism (Laidlaw 2014: 53).

For Faubion (2001), the potential of Foucault's work to inspire theoretical and empirical anthropological advancements stands in his understanding, adoption, and extension of a truly Aristotelian approach to practice in which he invites us "to attend to all that distinguishes the ethical field beyond mere obedience (or mere transgression)" (Faubion 2001: 85). In other words, as Moore (2011) has also argued, an analytic focus on ethical practices can surpass a view on agency as being solely expressed in relation to structure. Foucault makes an important distinction between moral codes and ethics. Moral codes—or what Faubion (2011) calls the themitical—are sets of rules or regulations enforced by various institutions, while ethics consist of "the ways individuals might take themselves as the object of reflective action, adopting voluntary practices to shape and transform themselves in various ways" (Laidlaw 2014: 111). While moral codes are cultural and social and they can be shared (Foucault 2000: 291), ethical practice is "always analytically distinct from the moral principles and codes to which it has reference" (Faubion 2001: 85). The ethical field is, therefore, a field of inquiry, of problematization that is not concerned with the question of obedience or transgression, but with doing reflective self-formation. In this view, change, which is both individual and social at the same time—as "individual" and "society," or concern about the self and concern about the others, are inseparable in Foucault's Aristotelian approach (Laidlaw 2014: 115)—comes from doing rather than from making.

Foucault identifies four parameters of the ethical domain: ethical substance, the mode of subjectivation, ethical work, and telos.

Ethical substance is the part of oneself that is the object of reflective consideration and work. Looking at ancient Greece in his genealogy of sexuality (1990), Foucault focuses on carnal pleasures as a form of ethical substance. Other parts of oneself that have been regarded as ethical substance in various cultural and historical contexts are the soul, one's identity, and certain kinds of acts and desires. Ethical substance is approached through the impulse of curiosity. The type of curiosity advocated by Foucault is "not the curiosity that seeks to assimilate what is proper for one to know, but that which enables one to get free of oneself" (Foucault 1990: 8). From this perspective, curiosity evokes care: "it evokes the care one takes of what exists and what might exist; a sharpened sense of reality, but one that is never immobilized before it; a readiness to find what surrounds us strange and odd; a certain determination to throw off familiar ways of thought and to look at the same things in a different way" (Foucault 2000: 325).

The mode of subjectivation represents the way in which people position themselves in relation to their ideals, values, and to the rules they are following: "the way in which the individual establishes his relation to the rule and recognizes himself as obligated to put it into practice" (Foucault 1990: 27). While Foucault intends this parameter to be "an index of the 'deontological'—precisely that aspect of the ethical domain which has to do with obligation or duty" (Faubion 2011: 50), his conceptualization leaves room for a range of ways other than duty in which people might relate to their ideals (Laidlaw 2014: 103). The mode of subjectivation can be regarded as self-stylization in relation to social norms and in relation to one's ideals and values. The process of choosing the mode of subjectivation that would accompany one's ethical project is not merely an act of conformity in which one needs to pick the mode that would suit them better from a variety of socially accepted and available modes, but it is an active and creative process. Drawing from Baudelaire's view of "modernity" as an attitude, Foucault sees self-stylization as "an exercise in which extreme attention to what is real is confronted with the practice of a liberty that simultaneously respects this reality and violates it" (Foucault 2000: 311). In the creative process of choosing and making one's own mode of subjectivation, there is social critique. As Faubion remarks, the ethical field is a normative field, but there is more to it than that: "It is a domain of obedience. Yet it is also a domain of more elective aspirations, of the 'quest for excellence, as we moderns like to put it, of saintly and heroic excess" (Faubion 2001: 90).

Ethical work, or *askêsis*, is the work that one performs in order to "attempt to transform oneself into the ethical subject of one's behaviour" (Foucault 1990: 27). This is a work of reflection that can combine diverse techniques and activities or "technologies of the self" (Foucault 2000: 223–251) that are available in a particular cultural and historical context in various personal or individual patterns. Some forms of technologies of the self that were available in ancient Greece were physical exercises, diary keeping, the interpretation of dreams, and the writing and exchange of personal letters. The thought and the self-reflexivity involved in ethical work are seen as "freedom in relation to what one does, the motion by which one detaches oneself from it, establishes it as an object, and reflects on it as a problem" (Foucault 2000: 117). Ethical work is essentially practical—it is a form of doing, not of making, in Aristotle's distinction—and, in Faubion's reading, it can be regarded as an "autodidactic" endeavour (2001: 94).

Finally, telos represents the mode of being, or the kind of being, that the ethical subject aspires to be: "A moral action tends toward its own accomplishment; but it also aims beyond the latter, to the establishing of a moral conduct that commits an individual, not only to other actions always in conformity with values and rules, but to a certain mode of being, a mode of being characteristic of the ethical subject" (Foucault 1990:

28). Rabinow (2000) suggests that the mode of being that Foucault was committed to was one of disassembling the self, or releasing oneself from oneself (*se déprendre de soi-même*), which could also be translated as unlearning, or de-familiarizing with, oneself (as in, I would add, the process of de-familiarization required by ethnographic work). Foucault regards the ethical category of mode of being as historically constrained but also as having the capacity to surpass the present by engaging with it critically and through de-familiarization. Moore expresses this idea suggesting that "how we are placed in time and space links to modes of being–specific ways of thinking, feeling and acting, of relating to things, to others and to ourselves. Political and economic changes alter these ways of being, and new ways of seeing and understanding drive forward further possibilities for change" (2011: 1–2). These new ways of seeing and understanding emerge, in Foucault's opinion, through ethical practices— through the process of ethical analysis as a practice of freedom.

Ethical work can also be regarded as a practice of "problematization" (Foucault 1990: 10). For Faubion, problematization is "that reflexive process through which one presents to oneself a certain way of acting or reacting, asks questions of it, examines its meanings and goals" (2001: 97). He argues that Foucault's work on the genealogy of systems of thought demonstrates that each parameter of the ethical fourfold–ethical substance, mode of subjectivation, ethical work, and telos–"can itself be (and in fact has been) the focus of problematization" (2001: 98). In other words, the four categories of ethical analysis have also been the subjects of ethical analysis; therefore, it is a never-ending process of self-reflection that drives social transformation. Faubion concludes that "the ethical field, in which power is fluid and problematization capable of being the catalyst of revisionary resolution, is the primary site of the active transformation at once of the parameters of subjectivation and of given views of the world" (2001: 99).

This conclusion takes us back to Moore's conceptualization of ethical imagination as one of the main sites of cultural invention (2011: 16). However, Moore considerably advances Foucault's ideas by insisting that ethical work, or the process of problematization, should be regarded as being more than a work of conscious self-reflection and thought. For her, problematization also involves "affect, emotion, the placement of the body, fantasy, and relations with objects, technologies and the material world" (Moore 2011: 21). Thence, Moore's focus on hopes, desires, and satisfactions aims to (re)introduce a regard for the value of imagination and for what I call the "non-conscious"—an experiential area that involves unintentional actions and the tacit knowledge of the body and of the unconscious—in processes of self-development or subject-formation. She argues "we cannot unproblematically base our theories of subjectivity on the experience of the conscious subject or on their conscious meanings and intentions" (Moore 2011: 75).

Foucault's work on ethics and ethical practices inspired the geographer Kersty Hobson to propose a new approach to everyday domestic practices in research focused on sustainability. In a recent publication, Hobson (2011) asks whether small, everyday acts of sustainability, such as the recycling of domestic waste, have the potential to make people more environmentally aware and active in the future, or if this approach to sustainable consumption can, on the contrary, limit the scope and the scale of the possible environmental engagements. Questioning the assumptions of several theoretical approaches to environmental policy interventions, Hobson suggests that researchers should not abandon the idea that a form of personal environmental politics might exist, because, in this way, they could play a key role in "not shutting down the debate on household sustainable consumption before it has got really interesting" (2011: 206). She proposes an approach inspired by Foucault's conceptualization of ethical domain, specifically looking at the question of "What do I aspire to be?" or the telos. Hobson argues that this question proved to be an important trigger in the way the people who took part in her research conducted in Australia reclaimed their streets as places of belonging—by bringing sofas in their front gardens and removing the fences—to express the fact that what they valued in everyday life was time and communication with others (Hobson 2008).

Drawing from Dikeç (2005), Hobson advocates for a more inclusive definition of the political, as appearing in "moments of interruption" when "a wrong can be addressed and equality can be demonstrated" (Hobson 2011: 203). In her research on everyday domestic practices in relation to the sustainability agenda in the UK and in Australia, Hobson (2003) identifies such moments of interruption as "ah ha" moments that her participants reported to have experienced when they made connections between everyday actions that they have not previously reflected upon, and their environmental outcomes. After such ah ha moments her participants believed that they could consciously work towards changing some of their domestic actions and reported a feeling of "I can do that, it's not so hard" (Hobson 2011: 204). These ah ha moments, as forms of everyday reflection, can be regarded as instants of ethical work in Foucault's terminology.

Hobson acknowledges that moments of interruption appear and are experienced as part of a wider existing grammar of discourses of accepted and positively valued practices and attitudes: "the range of available personal-political actions is circumscribed by discourses of accepted and acknowledged praxis, outside of which their impacts and meanings are lost–and indeed outside of which the ability to be labelled as political becomes problematic" (2011: 205). However, she suggests that existing grammars can be mobilized and shifted in order to make room for everyday ah ha moments to be acknowledged as political. This shift can start with scholars undertaking research on domestic sustainability—who can be regarded as positioned between environmental policy interventions and the everyday actions of laypeople—choosing to acknowledge such moments of interruption as part of a personal, creative and thoroughly political grammar of household sustainable consumption.

MUNDANE ACTIONS AND THE ETHICAL FIELD

In the previous two sections I have drawn a parallel between the work of the French philosopher Michel Foucault on ethical practice and Moore's (2011) post-vitalist call for a new acknowledgment of human agency and creativity. Foucault's conceptualization of the ethical domain and, specifically, of the telos also inspired the geographer Kersty Hobson (2011) to suggest that scholars researching domestic sustainability should be able to create a conceptual space in which the idea of a personal environmental politics could be recognized and developed.

While coming from different backgrounds and having different aims, I believe that Moore's and Hobson's calls are similar in spirit in at least two ways. First, they both urge social scientists not to abandon the belief in "the radical potentialities of human agency and human subjectivity" (Moore 2011: 22) in the ways they create theory and in the ways they regard, frame, and interpret the actions of their research participants. Second, they both try to open up the category of human actions that is considered to be worthy of attention and research to include ordinary and non-conscious actions, actions that have no conscious meanings or intentions attributed to them.

Their definition of worthiness is not related to the immediate possible effects on society that these actions could have but to the effects that they have on the agent. By adopting Foucault's conceptualization of the ethical domain, any and all types of everyday actions could be understood as moments of ethical work, that is, practical endeavours that contain their scope in themselves in the Aristotelian sense. Ethical work, or critical activity, is a practice of subject formation that is ongoing and that can be regarded, following Moore (2011) and Faubion (2001), as one of the main vehicles of social transformation. Ah ha moments provide the opportunity to critically re-examine the world and one's relationships with oneself and with others. As this re-examination—which does not need to be related to conscious meanings but can emerge from imagination and emotional and sensory experiences (Moore 2011: 75)—leads to a change in oneself, it would also lead to a change in one's relationships with others and, ultimately, to social change.

Looking at the conceptualization of the ethical in Greek antiquity, Foucault identifies the twin principles of "Take care of yourself" and "Know thyself." He argues that "our philosophical tradition has overemphasized the latter and forgotten the former" (Foucault 2000: 226) for two reasons. The first reason is that the tradition of Christian morality "insists that the self is that which one can reject" (2000: 228), the scope of knowing oneself being to provide a means of self-renunciation as the condition for salvation. The second reason is that, in theoretical philosophy, knowledge of the self is considered to be the first step in the theory of knowledge. Foucault argues that the importance attached to the two principles has been inverted. While in Greco-Roman antiquity knowledge of oneself was a consequence of the care of the self, in the modern world knowledge of oneself represents the fundamental principle. Foucault explains that care of the self was seen as having not only individual outcomes, but also important social outcomes: "the postulate of this whole morality was that a person who took proper care of himself would, by the same token, be able to conduct himself properly in relation to others and for others" (2000: 287).

In a study of grocery shopping in London, Miller (2001a) argues that decisions of buying organic products are very often related to concerns for the health of the shopper and of his or her family rather than to concerns for the world at large. This form of interpretation that aims to find the core reasons behind people's environmental concerns and actions can be found in other publications on ethical consumption as well.[2] The assumptions upon which scholars might compare two different sets of reasons, such as in Miller's case, could be seen, by following Foucault's (2000) argument, as grounded in the tradition of Christian morality and in a body/mind dualism. By acknowledging acts of taking care of oneself as essentially ethical and not as opposed to, but as the very basis of acts of taking care of others, one can surpass a set of analytical limits when situating everyday actions in relation to ideas about sustainability. This perspective would make it possible to re-examine a whole series of banal actions that social sciences scholars have explained by placing them in specific categories, such as the performance of social roles (e.g., interpreting mothers' expressions of care towards their children as performances of the role of mother) as ethical practices that are vehicles for social transformation.

Thus, for Moore (2011) and Hobson (2011) following Foucault, ordinary actions need to be acknowledged in themselves—and not only in relation to societal ideals and norms—because they have an effect on the agent. They do not affect the whole world directly, but by triggering critical activity and problematization, they are at the very basis of future social transformation.

Opening the domain of the ordinary to a fresh examination would ultimately bring about the possibility of starting to move towards a new vision of social change, a vision that in Moore's opinion would also contribute to social change: "We all clearly recognize that the way we think about things has an impact on the way we live in the world. Thought is a form of agency. Worlds and their futures are created by the actions of human beings in specific contexts, and representation is an indissoluble and crucial aspect of those actions" (2011: 143). Here, Moore points out to the responsibility that scholars have when they choose how to interpret their data, when they choose what story to tell to the others about what they have found out from their research. If one could claim that people's everyday choices have an impact on carbon emissions and climate change, then I argue that researchers' choices of theory to interpret their findings may be regarded as having carbon footprints of their own. Theoretical choices, like any other choices, have consequences.

NEOLIBERAL AGENCY AND ORDINARY AGENCIES

At this point in my argument, I introduce Ilana Gershon's (2011) critique of "neoliberal agency." In her essay, published in a special debate section of the journal *Current Anthropology* and followed by comments and a reply, Gershon addresses the challenges that neoliberal perspectives on agency pose to social anthropological interpretations of human action.

Gershon identifies two shifts that affect the concept of agency in the move from economic liberalism to neoliberalism. The first shift in a neoliberal context is that markets, economic rationality, and subjects are recognized as being made, achieved, or socially constructed. Gershon is inspired by the work of Wendy Brown (2003, 2006) and by Lemke's (2001) reading of Foucault's lectures on neoliberal governmentality, both of whom identify a major distinction between neoliberalism and liberalism as "the neoliberal emphasis on market rationality as an achieved state" (Gershon 2011: 538). In Gershon's opinion, the fact that neoliberal perspectives have incorporated a belief in social construction as an underlying principle that organizes the social would make an uncritical continuation of the analytical path of social constructivism in anthropology problematic. This is because, by dropping the concept of culture out of their analytical toolkit, social constructivist anthropological approaches do not have the capacity to oppose resistance to the fact that "a neoliberal perspective is additionally prescriptive by working toward universalizing forms of neoliberal agency" (2011:539).

While Brown (2003, 2006) and Lemke (2001) address the implications of the belief in construction for political practices, Gershon examines in her article the ethical implications of doing ethnography in neoliberal contexts. She argues that, for scholars coming from neoliberal contexts, it is increasingly difficult to find alternative, non-neoliberal ways of situating human action: "Spreading neoliberalism entails convincing others that everyone should enact corporate form of agency, produced by consciously using a means-ends calculus that balances alliances, responsibility and risk. Other forms of agency are getting pushed aside" (Gershon 2011: 539). Gershon argues that anthropologists have unintentionally contributed to spreading the neoliberal conception of agency by discarding the concept of culture in the last two decades. This concept could have provided a viable analytical apparatus (Wagner 1981) for revealing practical alternatives to neoliberal agency, and, while rejecting it, scholars did not find a different analytical tool to do this work with either. However, Gershon is clear in showing that the scope of her article is not to bring the concept of culture back but to draw attention to alternative tools that could support the analytical work that was once accomplished through the use of the concept of culture.

The second shift Gershon identifies that affects conceptions of agency is "a move from the liberal vision of people owning themselves as though they were property to a neoliberal vision of people owning themselves as though they were a business" (Gershon 2011: 539). She looks at the implications of this shift for a neoliberal conception of agency by discussing neoliberal perspectives on the self, relations, and social organization. Following this shift of vision towards the idea that people own themselves as though they were a business, the self comes to be regarded as an autonomous collection of skills, as Martin (2000) and Leve (2011) have also suggested. Neoliberal relations are alliances between different collections of skills, while social organization emerges "as market strategies that determine the consequences of alliances between different sizes of corporate entities and different skill sets" (Gershon 2011: 542). No longer part of the analytical toolkit of the researcher, culture is now regarded as a possession or a trait that can support specific sets of alliances.

Therefore, in a neoliberal context, agency is now simply and universally regarded as "conscious choices that balance alliances, responsibility, and risk using a means-ends calculus" (2011: 540). Agents are, therefore, assumed to be rational calculators who follow the principles of market rationality in the process of assessing the best suited strategies to create and to interpret relationships and forms of social organization. Departing from Comaroff and Comaroff's (2001) critique of neoliberalism as framing freedom only in terms of choice, Gershon argues that "it might be more apt to say that neoliberalism equates freedom with the ability to act on one's own calculations" (2011: 540).

In this situation, Gershon asks what the analytical techniques are that anthropologists could use to critique, rather than to contribute to spreading, neoliberal views of agency. In order to outline what makes anthropological approaches different from other social sciences' disciplinary approaches, Gershon discusses what an anthropological imagination entails in comparison to a sociological imagination. Drawing on Mills' (1959) definition of sociological imagination, she outlines that a sociological imagination "crosses levels of scale to produce insight, interrogating how practices at different levels of scale affect each other" (Gershon 2011: 543). This definition reflects the sociological endeavour of explaining how the individual is connected to the society and how the personal is connected to the political. For an anthropological imagination, however, "the challenge is to understand how the ways that people engage with knowledge and engage with each other are always already interwoven" (2011: 543). In other words, the anthropological endeavour is to understand how epistemologies and relationships shape each other. This process of mutual shaping, however, is not universal; it follows different paradigms and patterns for different groups of people in different places and in different moments of time. In Gershon's view, a focus on the mutual shaping between knowledge and relationships can be translated into an endeavour of employing the analytical tools of epistemological difference and social organization together. Even if these analytical tools might be seen as too closely tied to the concept of culture, Gershon argues that they could provide a fruitful critique of "neoliberal agency" by making possible a description of other ways of living and other contrasting and opposing conceptions of agency. Moreover, through sophisticated attention, social anthropologists would be able to attend to "the moral force of different epistemologies and social organizations" (2011: 547), which means more than mere description. It means also recognition of the fact that, by understanding alternative ways of engagement with the world, one comes to see paths to social transformation that go beyond neoliberalism.

I retain two important points from Gershon's essay that I will link with the literature that was discussed in the previous sections. First, the responsibility of scholars to choose their analytical tools and theoretical frameworks for interpreting data is brought to the fore in a strong and bold manner in this publication. As Amy Cohen, one of the essay's commentators outlines, "[a]nalytical tools have political power" (Gershon 2011: 547). The recognition of this fact, which appears as well in the works of Moore (2011) and Hobson (2011), is a central concern in Gershon's. The second point is the focus on agency as a domain that is particularly vulnerable in the face of political interpretations. By universalizing a specific definition of agency, neoliberal perspectives endorse a singular worldview. Neoliberal agency, I would add, does not account for people's capacity to produce radical social transformation, but only for their ability to rationally operate means-ends calculations within a given (neoliberal) context.

While Gershon calls for a renewed focus on social organization and epistemological difference to arrive at alternative descriptions of agency, Moore (2011) draws attention to epistemologies of the unconscious and imagination. I will now bring together Gershon's and Moore's ideas by proposing the working concept of "ordinary agency." Ordinary agency refers to forms of agency that are employed in everyday life and are related to non-conscious—unconscious, tacit, and non-intentional—acts of creating meaning. I place this working concept in contradistinction to the concept of "neoliberal agency"—as consisting of acts of conscious rational decision making—developed by Gershon, whose critique inspired me to look at the forms of agency articulated in the mundane. This working concept has two functions. First, it responds to Moore's (2011) and

Hobson's (2011) calls for an approach to ordinary actions as meaningful for social transformation. Second, it responds to Gershon's (2011) call for moving beyond "neoliberal agency" through attention to epistemological difference and social organization. I suggest that the domain of the ordinary, even in neoliberal economic and political contexts, can be regarded as manifesting different epistemologies—related to tacit knowledge, imagination and the unconscious—and bringing about new understandings of the mutual shaping between epistemologies and relationships.

To link this discussion back to the introduction of Foucault's writings on ethical work, I bring to the fore the focus on the practical. In Faubion's (2001) reading, the potential of Foucault's work to inspire epistemological advancements in anthropology stands in his continuation of Aristotle's approach to praxis, which could be seen as a way of surpassing a regard for human action as having either an obedient or a transgressive direction. In his definition of ethical work, or critical activity, Foucault draws upon and extends the Aristotelian distinction between making and doing, which suggests that while the end of making is different from it, doing contains its end in itself: "doing well is in itself the end" (Aristotle 1934: 337). By taking into account this distinction, one could suggest that "neoliberal agency" is concerned with making: making selves as collections of skills that are always extendable, making alliances, and making means-ends calculations. Neoliberal agency refers to action that always has a scope or a goal and can be easily identified and verbalized. Ordinary agency, instead, is concerned with doing. It would not be right to affirm that it refers to action that has no scope, but rather I suggest that the scope of this type of action (such as the wish to let the light come inside one's home every morning) is circumscribed within a domain of tacit knowledge. Therefore, it is not verbalized or even regarded as a scope every time the action is performed. Accordingly, I propose to regard ordinary agency as concerned with doing where doing is an end in itself.

AGENCY AND TIME

I will now introduce and discuss the last element that contributes to the theoretical toolkit of this work—the element of time and its connections to human agency that are discussed in the work of the anthropologist Carol Greenhouse.

Drawing from an interest in the relationships between law and time, which is related to her work in the field of anthropology of law, Greenhouse

(1996) makes a thorough analysis of time politics across cultures. One of the main arguments that she develops, and that I find to be of great significance here, is the idea that cultural models of time are inextricably linked to cultural understandings of agency, because "time articulates people's understandings of agency: literally, what makes things happen and what makes acts relevant in relation to social experience, however conceived" (Greenhouse 1996: 1). Greenhouse regards agency as a cultural concept because it names culturally situated ideas about "how the universe works" (1996: 4), allowing researchers to see what social relevance is in various social contexts. The main premise of her book is the idea that time is cultural rather than universal. Importantly, her take on time differs from Gell's (1992), whose reading of a similar body of scholarship has produced very different conclusions, because it was informed by a different premise.

Greenhouse suggests that most aspects of anthropological studies of social time have proceeded from the double assumption that linear time is the time of the Western world and time, as physical scientists conceive, is indeed linear—an idea that is reinforced in the temporal discourses of state nationalism. Her critical reading of the ways in which time was initially employed in social anthropological analysis, such as in the works of Evans-Pritchard (1940), Fortes (1949), Leach (1961), Lévi-Strauss (1969), Geertz (1973) and Bloch (1977), develops the idea that an approach to time as dual was often presented in universalistic terms. The duality of time concerned two geometrical forms: linear time, which was considered to represent social time and employed to understand progress and sequence, and cyclical time, which was regarded as being related to the periodicities of nature and enacted in the performance of rituals. Out of these two geometric forms of time, Greenhouse argues that linear time "is generally taken to be the more mundane, profane, practical, or objectively 'true' construction of time" (1996: 34). This idea, she shows, is linked to the assumption that the principal meaning of time comes from what is taken to be the universal significance of mortality. Therefore, the approach to linear time as being objectively true and universal comes from the fact that linearity is constructed in relation to mortality and, more generally, to the biological lifespan of human beings. By using a variety of ethnographic examples, however, Greenhouse suggests that death is not a universal object of individual preoccupation the anticipation of which creates a linear temporality. Rather, in several cultural contexts, it primarily represents a social crisis, an intrusive rupture of social solidarity instead of an individual preoccupation: "death is not about the closure of an interval but the disruption of the group and a corresponding crisis in social health" (Greenhouse 1996: 38).

The alternative to the duality of linear and cyclical time in ethnographic writing, she further shows, is a depiction of "timelessness." Some situations where a temporal dimension is unimportant or even lacking are, for example, the ways in which the Ilongot instead employ space as an element to organize the narrative of events (what we call history) (Rosaldo 1980) or the ways in which time and space are entangled in aboriginal Australian "dreamtime"(Bell 1993). Greenhouse argues that these examples, among others, show that "what makes time temporal is not necessarily an arrangement of past and future" (1996: 46) in relation to human mortality and to the periodicities of nature, but any number of other criteria that can be related to other formulations of social experience such as agency, space, and narrative.

In order to acknowledge the connections between time and human agency, we need to move beyond defining agency solely in relation to structure. We need to develop a much wider and more complex approach that recognizes understandings of agency as being linked to how people living in various contexts believe the universe works. For Greenhouse, agency "is not necessarily at the level of conscious choice or ideological commitment" (1996: 82) and needs to be depicted in much more diverse ways than by exclusively analysing its effects. She asks, "Can we consider agency as an ethnographic question without assuming in advance that what 'matter' about agency are individual effects on society or that what matter about time are 'moments'?" (1996: 81).

With this question, it is evident Greenhouse introduces an idea that is shared by the works of Moore (2011) and of Gershon (2011): the endeavour of decoupling agency and structure and moving beyond the assumption that agency only emerges as conscious choice. Gershon (2011) criticizes neoliberal agency as a way of universally defining agency as nothing more than a form of intentional calculation, while Moore (2011) called for an acknowledgement of unconscious forms of agency that are not directly concerned with effects on society. In the previous section I brought together Moore's and Gershon's ideas to formulate the working concept of ordinary agency, which describes forms of agency employed in everyday life that are related to non-conscious (unconscious, tacit, and non-intentional) acts of creating meaning.

However, what is specific to Greenhouse's approach is the focus on the relationship between time and agency. Her thesis is that cultural models of time are propositions about the nature of agency and the ways in which agency is distributed across social space. Looking at the temporalities of law in the USA, she argues that "linear time does the state's (or any institution's) work by providing an idiom with which individual agency can be represented as flowing into the nation's" (Greenhouse 1996: 180). She advocates a plural perspective that recognizes that individuals are able to simultaneously engage with multiple temporalities and to imagine and enact multiple forms of agency. However, Greenhouse remarks that the multiplicity of cultural formulations of agency "are only selectively acknowledged and accommodated by any particular institutional form" (1996: 233). An important role of the ethnographic endeavour here would be to describe and conceptualize some of the multiple formulations of agency that are not necessarily recognized as agencies by dominant institutional forms. Some examples of alternative formulations of agency in the USA mentioned by Greenhouse are creationism, laziness, superstition, and witchcraft—none of which are generally taken seriously in public discourse. She argues that these alternative formulations "should not be left as caricatures but read as critical, if partial, counter discourses of time" (1996: 7).

Temporal Modalities of Domesticity as Counterdiscourses of Time

Temporalities of domesticity, I believe, could also be regarded as counterdiscourses of time that make a consistent, even if quiet or non-conscious, statement about the ingrained presence of agency in everyday life. In the following chapters, I will describe and conceptualize three temporal modalities of domesticity—spontaneity, anticipation, and "family time"—together with the forms of ordinary agency that they engender and make visible.

For Greenhouse (1996), culture is an essential tool for arguing against the assumed universality of linear time and for investigating the plurality of conceptualizations of agency and time. As I have shown earlier, Gershon (2011) also argues that the disposal of the concept of culture left scholars with no analytical tools sharp enough to oppose neoliberal universalizing claims. In this work, which is based on my experiences as a Romanian anthropologist coming to the UK for the first time to conduct doctoral fieldwork, I want to acknowledge the fact that a dimension of cultural specificity might be intrinsic in the space and time that the people I met in Middleborough occupy and participate in. I understand this cultural dimension by adopting Moore's definition of culture as "an 'art of living', as a means of engagement with the world" (2011: 11). Therefore, the time modes that will be described in the next chapters can be regarded as connected to existing (and changing) forms of engagement with the world that are embedded, recognized, and performed in the English provincial and preponderantly middle-class universe where I carried out my fieldwork.

In such an environment, as Strathern (1992) suggests, a particular view on time is enacted in kinship relations. This is a concept of progressive time that flows downwards in relation to the irreversible succession of generations. The irreversibility of time is most visible here, Strathern outlines, in the relationship between parents and child where substance is passed from the former to the latter in an exclusively forward direction. "Causes thus flow forward in time. Consequently a person may be regarded as influenced by many things that happened before he or she was born, for he or she is born into a world already full of events and relationships. Parents affect children's identity much more than children affect parents. This downward or forward flow in time recurs as a question of individual development" (1992: 67). This reflection on the English sense of time in relation to kinship comes in the context of Strathern's analytical endeavour to define Victorian middle-class kin constructs and British anthropological kinship theory as "mutual perspectives on each other's modernisms" (1992: 8). She identifies three main facts of English kinship: the individuality of persons, diversity, and the idea that individuals reproduce individuals. Her observation about progressive time is explained in relation to the facts of English kinship where a child is regarded as an unique individual that appears "as an outcome of the acts of other individuals" (1992: 53). Time, therefore, is considered to increase the number of persons as well as, more generally, "the diversity and plurality of all the things in the world" (1992: 60).

In Chaps. 7 and 8, I will go back and discuss in more detail some ideas from Strathen's work on English kinship, especially the relationship between convention, morality, and having a "family-style lifestyle." For the moment, I take her observation on the sense of time in kinship as a point of departure for my analysis of the ways in which families approach, experience, represent, and enact domestic time. Therefore, if temporal direction is fixed to a forward flow, and if people cannot influence time's direction as they cannot interfere with the past either, then what can be influenced is the speed of the flow of time. In order to engage with and change time's gears, one might juggle with ideas of the present and the future in various ways. I propose to regard the concepts of spontaneity, anticipation, and family time as time modes of domesticity that play with, blend, and knead the present and future in various ways, and in so doing they constitute themselves as counterdiscourses of time in relation to linear time. Spontaneity and anticipation make the future present by creating wished-for future moments and making them happen and by coating the present with the imagination of the future, respectively. The intent of family time is to make the present future by prolonging it, by operating in the present continuous mode.

NOTES

1. See also (Miller 2001b).
2. See (Morosanu 2013) for a review.

REFERENCES

Aristotle. 1934. *The Nicomachean Ethics.* London: William Heinemann Ltd.

Bell, Diane. 1993. *Daughters of the Dreaming.* Minneapolis, MN: University of Minnesota Press.

Bloch, Maurice. 1977. "The Past and the Present in the Present." *Man* 12 (2): 278–292.

Brown, Wendy. 2003. "Neo-Liberalism and the End of Liberal Democracy." *Theory and Event* 7: 1.

Brown, Wendy. 2006. "American Nightmare." *Political Theory* 34: 690–714.

Comaroff, Jean, and John Comaroff. 2001. "Millennial Capitalism: First Thoughts on a Second Coming." In *Millennial Capitalism and the Culture of Neoliberalism,* edited by Jean Comaroff and John Comaroff, 1–56. Durham, NC: Duke University Press.

Das, Veena. 2007. *Life and Words: Violence and the Descent into the Ordinary.* Berkeley, CA: University of California Press.

Dikeç, Mustafa. 2005. "Space, Politics, and the Political." *Environment and Planning D: Society and Space* 23 (2): 171–188.

Evans-Pritchard, E.E. 1940. *The Nuer.* Oxford: Oxford University Press.

Faubion, James D. 2001. "Toward an Anthropology of Ethics : Foucault and the Pedagogies of Autopoiesis." *Representations* 74 (1): 83–104.

Faubion, James D. 2011. *An Anthropology of Ethics.* Cambridge: Cambridge University Press.

Fortes, Meyer. 1949. "Time and Social Structure: An Ashanti Case Study." In *Social Structure: Essays Presented to A. R. Radcliffe-Brown*, edited by Meyer Fortes, 54–84. Oxford: Clarendon Press.

Foucault, Michel. 1990. *The History of Sexuality*, vol. 2. London: Penguin Books.

Foucault, Michel. 2000. *Ethics: Subjectiviy and Truth (Essential Works of Foucault 1954–1984)*. London: Penguin Books.

Geertz, Clifford. 1973. *The Interpretation of Cultures: Selected Essays*. New York: Basic Books.

Gell, Alfred. 1992. *The Anthropology of Time: Cultural Constructions of Temporal Maps and Images*. Oxford: Berg.

Gershon, Ilana. 2011. "Neoliberal Agency." *Current Anthropology* 52 (4): 537–555.

Greenhouse, Carol J. 1996. *A Moment's Notice: Time Politics Across Cultures*. New York: Cornell University Press.

Hobson, Kersty. 2003. "Thinking Habits into Action : The Role of Knowledge and Process in Questioning Household Consumption Practices." *Local Environment* 8 (1): 95–112.

Hobson, Kersty. 2008. "Reasons to Be Cheerful: Thinking Sustainably in a (Climate) Changing World." *Geography Compass* 2 (1): 199–214.

Hobson, Kersty. 2011. "Environmental Politics, Green Governmentality and the Possibility of a 'Creative Grammar' for Domestic Sustainable Consumption." In *Material Geographies of Household Sustainability*, edited by Ruth Lane and Andrew Gorman-Murray, 193–210. Farnham, UK: Ashgate.

Laidlaw, James. 2014. *The Subject of Virtue: An Anthropology of Ethics and Freedom*. Cambridge: Cambridge University Press.

Lambek, M. 2008. "Value and Virtue." *Anthropological Theory* 8 (2): 133–157.

Latour, Bruno. 2005. *Reassembling the Social: An Introduction to Actor-Network Theory*. Oxford: Oxford University Press.

Leach, Edmund. 1961. *Rethinking Anthropology*. London: Athlone Press.

Lemke, Thomas. 2001. "'The Birth of Bio-Politics': Michel Foucault's Lecture at the College de France on Neo-Liberal Governmentality." *Economy and Society* 30: 190–207.

Leve, Lauren. 2011. "'Identity.'" *Current Anthropology* 52 (4): 513–535.

Lévi-Strauss, Claude. 1969. *The Raw and the Cooked: Introduction to a Science of Mythology*. New York: Harper and Row.

Martin, Emily. 2000. "Mind-Body Problems." *American Ethnologist* 27: 569–590.

Massumi, Brian. 2002. *Parables for the Virtual: Movement, Affect, Sensation*. Durham, NC: Duke University Press.

Miller, Daniel. 2001a. *The Dialectics of Shopping*. Chicago: University of Chicago Press.

Miller, Daniel. 2001b. "Possessions." In *Home Possessions: Material Culture behind Closed Doors*, edited by Daniel Miller, 107–122. Oxford: Berg.

Mills, Wright. 1959. *The Sociological Imagination*. Oxford: Oxford University Press.

Moore, Henrietta L. 2011. *Still Life: Hopes, Desires and Satisfactions*. Cambridge: Polity Press.

Morosanu, Roxana. 2013. "Ethical Consumption: Social Value and Economic Practice, by J. Carrier and P. Luetchford." *Anthropology in Action* 20 (1): 49–51.

Ortner, Sherry. 1984. "Theory in Anthropology since the Sixties." *Comparative Studies in Society and History* 26: 126–166.

Pink, Sarah. 2004. *Home Truths: Gender, Domestic Objects and Everyday Life*. Oxford: Berg.

Rabinow, Paul. 2000. "Introduction: The History of Systems of Thought." In *Ethics: Essential Works of Foucault 1954–1984*, edited by Michel Foucault. London: Penguin Books.

Rosaldo, Renato. 1980. *Ilongot Headhunting, 1883–1974: A Study in Society and History*. Stanford: Stanford University Press.

Strathern, Marilyn. 1992. *After Nature: English Kinship in the Late Twentieth Century*. Cambridge: Cambridge University Press.

Thrift, Nigel. 2008. *Non-Representational Theory: Space, Politics, Affect*. London: Routledge.

Wagner, Roy. 1981. *The Invention of Culture*. Chicago: University of Chicago Press.

CHAPTER 3

Encountering Middleborough: Impressions, Methods, and Tacit Knowledge

My work as part of the Low Effort Energy Demand Reduction (LEEDR) project involved a forty-two-month doctoral studentship that was concurrent with the duration of the project. This work also required me to take part in project meetings and other activities while I was developing my own research focus for my PhD. As the academic affiliation of the project and the field site were both located in Middleborough, accepting the doctoral position meant moving to the town for the entire duration of the research project.

The opportunity to spend over four years in the town where I carried out my doctoral fieldwork has proven to be fertile and life-changing, though my reaction after first visiting Middleborough[1] in September 2010 was less optimistic. It was my first time in the UK, where some of my favourite writers, filmmakers, and music bands originated, and I was expecting each street corner to display the quality of exciting and creative heterogeneity that provided inspiration for the people whose work I admired. This expectation was also related to a general predisposition in Romania to look up to Western countries as places that are more developed economically and where people are, generally, happier, nicer, and smarter. Coming from lively Bucharest, where I was living in the city centre surrounded by theatres and music venues, the perspective of spending the next four years of my life in a place that looked dim and uneventful, instead of conducting between twelve and eighteen months of fieldwork in a warm

© The Editor(s) (if applicable) and The Author(s) 2016   45
R. Moroşanu, *An Ethnography of Household Energy Demand in the UK*, DOI 10.1057/978-1-137-59341-2_3

climate and then continuing my old life as I thought that anthropologists "normally" did, felt like ostracism. After that first visit I asked my supervisors if it would be possible to spend part of my first year in London to attend open lectures at anthropology departments (and to experience the city vibe), do multi-sited fieldwork, or move back to Romania in the final eighteen months to "get distance" for the stage of writing my dissertation. The first two ideas did not materialize because of the constraints of working as part of a wider project that had a set research design in place and required my presence at fortnightly meetings and my contribution to various tasks. I forgot all about the third idea when I met my boyfriend Steven, a Middleborough resident, towards the end of my fieldwork.

As I write these words, my field site has (provisionally) become my home. I travel through its brisk quiet air every day and wonder how I could better describe Middleborough's overwhelming qualities of discreet smallness and persisting uniformity, which felt constraining in the beginning but familiar and nearly unnoticeable now. I pass by houses where my participants live in the evenings on my way home and try to determine whether they are home and what they are doing by the rooms where the light is on. By looking at the lit windows of the people I am writing about, I reassure myself that they are real—an everyday reassurance that is impossible for someone to reach with similar intensity when they leave the field site at the completion of the fieldwork.

I moved to Middleborough and I started my PhD in December 2010. During my first months here, living with my landlady Melanie who was also working towards a PhD at the local university, I was surprised by the fact that I could never see our neighbours. Melanie's three-bedroom house was in a suburban residential neighbourhood that was located a forty-minute walk from the town centre and a twenty-minute walk from the campus. While not too long and time consuming compared to commuting in a large city, the walks felt lonely because I was only rarely passing by other people walking on the pavement. The sound and sight of passing cars was, however, uninterrupted. I realized that the only time to see our neighbours, was in the moments after they got out of their houses and before they got in their cars—moments that I kept missing. Houses with lit windows, then, felt like inaccessible fortresses where people were hiding away and hibernating.

Springtime brought some changes. When a football landed in our back garden, I walked with the ball under my arm, full of hope, to the parallel back street and tried to determine which house it came from. I went to the house with the garden behind ours and rang the bell. A woman, two kids, and a dog all rushed to the door; they were glad to see the ball back, but they did not offer me a cup of tea and we did not introduce ourselves and shake hands as neighbours do in Romania. They told me not to bother next time and just throw the ball over the fence. I started to worry: how could I ever do fieldwork with English families when I was not able to meet them? In my home country, village forms of sociality in which everyone knows everyone else are reproduced in urban settings in the socialist flats—the exclusive form of urban dwelling. People usually know their neighbours, pay them regular visits, chat endlessly in front of the elevators, and generally know about and are concerned with their problems and circumstances. What was happening inside English houses? What forms of joyfullness, complicity, and harmony were achieved at home to make up for the isolation? I could not imagine an answer to these questions at the time nor to why someone would choose to live in Middleborough and why a young person would spend their degree years here.

Ten months later, however, when I started my "official" year of fieldwork, I was already part of a variety of local groups and networks. For example, I joined a Morris dancing group, the Transition Town Middleborough grassroots movement, and two other groups that organized fortnightly political discussions and meetings with council representatives on topics related to transport and urban planning in the area. I had English friends, and I started to develop an understanding of their reasons for living in Middleborough. After I moved closer to the town centre with a group of friends, where our cheerful neighbour was always chatting with us over the garden fence and was happy to try my home-brewed beer, I started to enjoy living in Middleborough much more. My relationship with the town was officialized when I found my photo and my name in the local paper: a few times it was a group photo of my Morris side, and another time it was a photo taken at a mending event on a weekday morning where me, an elderly friend, and his elderly lodger were the only participants. I bought several copies of that edition of the local weekly paper and sent them to my parents and my friends back home as proof that I was an acknowledged presence, a provisional member, an "anthropologist." In time, amazing things happened. I was in someone's house for the first time attending a spontaneous after party following one of the ceilidh group's weekly sessions and recognized someone I knew from a completely different context in a photo on the host's fridge. I was recognized at an open garden event by a lady who had seen me Morris dancing and who knew other members of my dance group. Saturday market trips in the town centre started to be accompanied by unplanned meetings with people I knew. Relationships and connections, like the streets on an emotional map, made the town alive and enjoyable for me to live in.

Once, when I attended a public meeting organized by a local group interested in improving public transport, the main characteristics of the town were discussed. An urban planner from the council read a list of main features from part of the development master plan for the town centre that, the local group members agreed, made the town unique: its pedestrian market place, its walkable human size, the market on Thursdays and Saturdays, the Georgian and Art Deco architecture, the mix of shops that represented both independent businesses and big brands, the green parks, the university and the international environment, and the Friday night culture. It was the mixture of these qualities, rather than any of these characteristics taken separately, that made Middleborough different from other towns.

While the population (50,000 to 70,000 inhabitants) is that of a medium-size town, Middleborough has a small-town feel because there are no other buildings, besides a few student halls, higher than three levels. However, I discovered that the impression of smallness was one of the things that Middleborough residents found attractive. The vast majority of the people I met here, as well as the majority of the participants in the LEEDR project, were not originally from Middleborough. They had moved from other parts of the UK, following jobs or available nursery schools and a direct rail connection to London. People often emphasized that (according to my field notes) Middleborough "is not too big, but big enough," that "it's a town, not a city," and that "you've got everything you need in a walking distance" while still being able to go for a walk in the woods straight from your back garden if you live on the forest side of town.

The way in which I developed the pseudonym Middleborough for my field site was, first, in relation to the cultural categories of "the city" and "the country" (Williams 1973) that the residents made use of in describing the town. In Williams's (1973) analysis of English literature, images of the country and the city appeared in opposition—the country represented a realm of simple and unadulterated life, while the city was depicted as a symbol of alienation and capitalist exploitation. Though the meanings associated with these two realms have evolved and changed over time, the city and the country are, arguably, still regarded as representing two essentially different ways of living in the UK.[2] However, I argue that Middleborough can be seen as providing the best of both worlds: it is neither a city nor a village, it is both. Living in Middleborough, people do not need to choose between a countryside lifestyle or a city lifestyle, between being a countryperson or a city person. Residents do not have to refuse the one when choosing the other. Rather, as an assemblage of both worlds, the tensions inherent in the city vs. country dichotomy are domesticated in Middleborough through an emphasis on the positive dimensions of both forms of living and a rejection of negative dimensions such as the alienation and anonymity of urban life and the normative nature of a village where local-born residents can be reluctant to welcome "newcomers" (Strathern 1981), "offcomers" (Rapport 1993), or "incomers" (Edwards 2000). I now use a similar idiom when I describe the place where I live to my friends in Romania. I rent an old Victorian terraced cottage in the town centre, and when I step outside of my house I can see the best pub in town, The Lemon Tree. I pass by two Chinese restaurants and a fish and chips shop on my way home. The main gym and leisure centre is a two-minute walk from the pub, and the university campus is within fifteen minutes. And Beacon hill, where some of the oldest stones in England can be found and where the local grove organizes May Day celebrations and Equinox and Solstice rituals, is a ten-minute car journey away.

An additional cultural dichotomy I wanted to reflect in the name Middleborough is a distinction between the categories of "the North" and "the South", or "people from the North" and "people from the South." I encountered this distinction many times during my fieldwork and during my life in England. For example, when my company found out that I was an anthropologist, I was sometimes asked what I thought about the differences between the North and the South. On other occasions, people told me that I was lucky to be a foreigner in the UK because I could travel to any part of the country and the locals would treat me nice, while they could not travel to an area located at the other end of the UK and feel welcome because they would be betrayed by their original accent. The awareness of differences between people from the North and people from the South, and an interest with constantly observing and comparing them, makes any English resident into something like an anthropologist. While these cultural differences have been situated historically (Jewell 1994), it was during the Thacherite government—when economic and political differences between these two regions (an affluent and Conservative South vs. a poor Labour-voting North) became more pronounced and visible—that the so-called North-South divide was acutely emphasized and discussed in the media, as Lewis and Townsend (1989) observe. Middleborough accommodates both those who moved from the North and those who moved from the South because of jobs, partners, or changes brought about by different stages of life. Those who come from the North prefer Middleborough because it is less rainy than the place they originated from, while those who come from the South appreciate that they have more routes for countryside walking here because a substantial part of the land is privately owned in the South. Middleborough, therefore, provides a critical distance for people from both the "rough North" and the "stately South."

By being neither on one side nor on the other of the "rough North" and "stately South" divide or the "cosy but normative village" and the "exciting but alienating city" dichotomy, Middleborough is liberating in the sense that it represents a context that no strong characteristics have been necessarily associated with. One could say that, perhaps in the imagination of an English resident, life in Middleborough might be seen as less prescriptive than life in one of the well-defined four poles that I discussed.

During the lengthy process of learning to understand the reasons why one would live in Middleborough, one of the experiences that most broadened my imagination and my intuition in tacit ways was becoming a Morris dancer. I met Zia and Richard in late spring during my first year in Middleborough after a performance organized by the university's contemporary arts programme that Zia and me volunteered for as performers. While chatting with Richard afterwards in a pub, he told me that they were starting up a Border Morris side. After discovering that I was an anthropologist and was going to be in Middleborough away from my family and friends for three more years, he insisted that I join them. He said that this would be a good way for me to learn about real English traditions and they could become my family for the next three years if I join the group. I agreed to go to their next practice, which was held on Sunday evening in a place called The Druid Arms that Richard gave me directions to. The next day I searched Morris dancing on Wikipedia and saw photos of people wearing white tattered costumes, which reminded me of a Romanian traditional ritual dance called Calusarii that the American anthropologist Gail Kligman (1981) wrote a monograph about. That evening, however, while watching a live TV comedy show, my attention was captured by a close-up of and short interview with a group of members of the audience who were wearing long hair, long beards, and long cloaks—similar to the cast of the film *The Lord of the Rings*—who mentioned that they were druids. I wondered if the place called The Druid Arms was the headquarters of the Middleborough druids and if I was signing up to become a member of a secret organization. I was relieved to discover after looking for more information about this place on the Internet that it was the name of a pub.[3]

I started attending the Morris dancing weekly practices, which later moved from the back room of the pub to a hall in a community centre, the following Sunday. Zia and Richard taught me the dance steps with great patience, and they or other members of the group always gave me a lift to and from practice. We grew very close in time and got to know one another in a way that perhaps only shared physical tasks permit. We did numerous public performances together and participated in folk and pagan festivals and events all across England where I was often the only non-British participant.

In addition to opening up a whole new social world for me, the experience generated a new relationship with my body. I have never been good at choreographed dance or synchronization, and after the first few practice sessions, I thought it would be impossible to get attuned with the other dancers. In time, by learning the tunes and getting to know the members of the group, by responding to the energy, complicity, and closeness resulting from a successfully accomplished dance, I developed the skill of clearing my thoughts when performing and letting my body respond to the music. One of the things that made Border Morris different from other forms of choreographed dance and helped me master the skill was the dialogic relationship with the music. In Border Morris, dancers use sticks that they clash during the chorus as a form of percussion: the sound of sticking supports the tune, integrates music and dance into a whole, and enhances the multi-sensory experience of performing the dance. After practicing and habituating a dance as a complex multi-sensory sequence of actions that displays a specific music-dance relationship, the body remembers and enacts the dance when hearing the music. Having mastered the dances, I now find it easy to learn new ones during gatherings with other sides; the structure and combination of sequences of the dances makes complete sense to me in a tacit way, like a new language that my body now knows.

Even if I had been interested, theoretically and academically, in tacit knowledge before, it was through Morris dancing that I really learned to trust my body, my intuition, and my emotions, as well as to acknowledge them as instruments of knowledge formation. Following the distinction between "knowing that" and "knowing how" (Ryle 1949)—or in Bloch's (1985) terms, ideological (propositional) and everyday (non-propositional) knowledge—Harris (2007) observes that ethnographers have usually dealt with the first category, "the explicit part of the information they gather in the field, such as what is told them, what they observe and can measure" (2007: 12), and have been less preoccupied with developing methodological practices and instruments that would tackle tacit forms of knowledge related to body techniques, to skills, and to sensory experience. Tacit knowledge, or what the body knows, is a form of everyday knowledge that people do not usually reflect upon. Thus, it is non-propositional: it appears in actions rather than in words.

While my ethnographic fieldwork with English families did not involve folk dancing, I believe that the skill of Morris dancing, which my body mastered as a form of tacit knowledge, opened up research questions that I was able to formulate and to address within the defined context of my official research. I approach the relationship between these two distinct parts of my ethnographic fieldwork—one related to everyday life and serendipitous experiences at my field site, the other related to my well-defined research of English families as part of the LEEDR project—through Grasseni's (2007) perspective on skilful vision. She argues that the process of coming to apprehend, or to "skilfully see," locally achieved understandings or worldviews that are embodied in a community of practice is essentially practical: "By participating in many different constellations of communities of practice, in various capacities and degrees in the course of one's life, one pragmatically gains the capacity to relate to those 'forms of life' that are in some way contiguous to one's own" (Grasseni 2007: 208). In other words, the wider the variety of activities one engages in while performing fieldwork, the further one's capacity to relate, to intuit, and to understand the worldviews of others develops. Morris dancing, therefore, was a form of bodily training for me, in addition to gardening, English cooking and baking, beer brewing, ceilidh dancing, countryside walking, and other activities. It helped me develop a sensitivity to other forms of understandings that my English participants tacitly held, such as understandings of domestic time.

The expressive characteristics of Border Morris dancing, like sticking and shouting, make this a particularly agentive activity. The dancers look and act self-confident and loud. We wear tattered jackets and an abundance of accessories: collections of badges - from Morris events or with the name and logo of other dance sides - that show one's experience and connections; hats decorated with ribbons and feathers; earrings, necklaces and rings; and long gloves and colourful knee-length socks. We paint our faces in full black, with stripes, such as those from popular culture representations of Native Americans, or with various pagan patterns. We wear straps of small bells tied on our ankles or under our knees, which signal every step we take by accompanying them with rings. Morris dancers are a conspicuous presence that people acknowledge with amusement. By privileging a loud, physically intense, and playfully aggressive expression, Morris dancing makes performers into agents. I believe that the constant sense of agency that I derived from dancing contributed to me shaping the focus of my thesis on the pursuit of identifying and defining alternative forms of agency that are not conscious and end-driven but that are ordinary and tacit.

THE DOUBLE BIND OF APPLIED AND ACADEMIC ANTHROPOLOGY

While my initial experience of the town was of a place of hidden and inaccessible inhabitants, my work life was populated by strong presences. I was engaged in frequent and lively interdisciplinary encounters with my colleagues working as part of LEEDR. The project team consisted of fifteen investigators and researchers from the disciplines of building engineering, electrical and systems engineering, computer science, design, and social sciences. During the initial fortnightly project meetings before we advertised for and recruited the participants, our dialogues focused on the pressing question of what sort of research we all wanted to do together. These meetings often turned into discussions about our disciplines - looking at epistemological and methodological beliefs, and opinions of what important and worthwhile research was - that could not have been more different from one another. I enjoyed these dialogues and found the challenge of explaining to peers coming from technical backgrounds what anthropology is and what it does to be a useful exercise in refining my own thinking about the ways in which knowledge gained from ethnographic research could be used to make the world better.

However, as I was the only representative of the social sciences branch of the project for the first seven months—my supervisor was away on sabbatical and a social sciences research associate was appointed only later—there were moments when the demand to constantly explain and legitimate my disciplinary approach felt daunting. For example, I was

asked to think about all the questions I could possibly address to participants during my fieldwork and how their answers would contribute to the
overall aim of the project. It was the first time that I was taking part in an
interdisciplinary academic project, and I soon found out this was a very
different approach to work than the one in commercial companies, which
I had a few years' experience in before embarking on a PhD. While team
meetings in a corporate environment are led by the team's coordinator or
manager and are always very short and focused on setting and distribut-
ing tasks to all the people in the room, the hierarchy in an English academic context is structured in much more complex ways. In an academic
context, some disciplines rather than one individual try to influence the
type of research that other disciplines plan to conduct, and agreement is
often more important than effectiveness. A note in my diary from the first
year of the project compares the structure of an academic interdisciplinary meeting with the structure of a musical, a performing arts genre that
I first encountered in the UK: both feel lengthy and the conflict is rather
minor compared to Greek tragedies. Different characters might have different opinions and talk about them at length, but everyone agrees in the
end and the social order is re-established. Holding strong views or, in
general, any form of social critique in an academic interdisciplinary meeting is unusual. It is sometimes regarded as a sign of non-cooperation or
of trying to be "weird" or difficult—just as musicals are not about social
change: the audience leaves the venue heart-warmed and comforted, not
pensive and uneasy.

During this initial phase of the project, I also travelled to London to
attend open lectures in anthropology as well as a postgraduate discussion
group on material culture. This time, when talking about my doctoral
research with peers, I discovered a different wall that our conversations
would invariably run into. This always happened when mentioning that
my research was part of an applied project in which the objective was to
develop interventions that would help reduce domestic energy consumption. While never due to explicit verbal reactions, after these exchanges
I was sometimes left with the feeling that my research was not as interesting, sophisticated, or "anthropological" as the research conducted by
other doctoral students who were working independently, did not have
any applied concerns to respond to, and were doing fieldwork outside
the UK. Looking at ways to respond to and improve a societal issue of
actual public concern in England was considered to be the endeavour of
sociologists rather than of anthropologists. The questions that informed

the overall research project part of that I was working on did not have enough anthropological sophistication, which meant that my own research would likely not be anthropological enough.

I found myself between these two positions of applied interdisciplinarity and academic anthropology, wanting to occupy them both while constantly feeling that through this very attempt I was not doing any of the two parts of my work well enough. My previous social anthropological training made me question the assumptions embedded in the funding call that the LEEDR project was a response to, which was apparently not exactly the type of "productive" contribution that the team members from other disciplines were expecting. Equivalently, while being animated by soft green political convictions and by the belief that reducing energy consumption in the global North was generally a good idea, I thought that social anthropology could play an important role in the articulation and mitigation of such change. I thought that collaborating with people coming from other disciplinary backgrounds should not be assumed to necessarily lead to a decrease in the level of depth, or "thickness," of the ethnographic description and theory that would emerge.

Anthropologists who have found themselves in a position of the "in between"—it is often the case of postgraduate-trained researchers who conduct applied research—have written about the relationships that their new status forms, both with traditional academia and with teammates and co-researchers from other backgrounds. Regarding the view from academia, it is argued that applied work may still be viewed as "impure" and compromised (Roberts 2006; Wright 2006; Sunderland and Denny 2007). When Sunderland and Deny started their own qualitative market research company and applied the ethnographic methods and social anthropological analysis that they had been academically trained to employ, their applied work was assumed, from the academic side, to be "less theoretical, less sophisticated, and ultimately less valuable" (Sunderland and Denny 2007: 31). In trying to explain this assumption that he has also often encountered, Roberts (2006) identified a series of elements that are anathema to academic anthropologists: "paid informants, the very short period of time spent with the households and a narrow focus of vision on the part of the research team" (2006: 84). Concerning paying informants, Drazin (2006) observes that for his research participants in the UK, this is an expectation and a proof that the research was "serious." He suggests that his British informants prefer "a commodity form of data," while the data in his previous research in Romania was constructed "in a stereotypical 'gift' form" (2006: 105).

Moreover, the cited scholars talk about a possible scale of increasing degrees of impurity in relation to the goal of the research: "if 'applied' in general was dirty, consumer research or 'marketing' was filthy—wickedly so, in fact" (Sunderland and Denny 2007: 31). This point is emphasized by the debate around the vocabulary used to describe applied anthropology. In a historical account on anthropological applications in policy and practice in the UK over two decades, Wright (2006) shows how the name of the Group for Anthropology in Policy and Practice (GAPP), part of the Association of Social Anthropologists of the UK and Commonwealth, was chosen in the 1980s so as to avoid the term applied anthropology, which "carried connotations of a separate profession, parasitical on 'pure' academics for ideas which they applied without making contributions to the development of the discipline themselves" (2006: 33).

A more recent debate in the US also focused on the relationship between applied and public anthropology. At Borofsky's (2000) call to academics for public anthropology engaged with wider audiences, Singer reacted by showing that this is what applied anthropologists have been doing for a considerable period of time. He suggests that Borofsky's appeal shows the "conscious non recognition of applied anthropology" (Singer 2000: 6), and he outlines that "given that many applied anthropologists already do the kinds of things that are now being described as PA [public anthropology], it is hard to understand why a new label is needed, except as a device for distancing public anthropologists from applied anthropology" (idem).

In my work as part of the LEEDR project, a series of unexpected benefits of applied interdisciplinary collaboration emerged in the stages of obtaining ethical approval and recruitment. Both of these processes showed that when working collectively rather than as a lone researcher, responsibilities are distributed and the gains coming from acknowledging and drawing upon the specializations of co-workers are considerable. While these processes were led by a colleague from building engineering, the vast majority of project members took part in several discussions about ethical issues and lone working procedures, and all the researchers went through a Criminal Record Bureau (CRB) check to be able to spend time with the family participants inside their homes.

Twenty home-owner families were recruited through advertisements in the local paper, school and university newsletters, and posters that were placed in community centres and general practitioner medical centres. The participants were not incentivized and they volunteered to take part in the research for three years for various reasons, from an interest in sustainability and low-carbon ways of living to a wish to lower their energy bills. With the exception of one single-parent family and one extended family where the maternal grandmother was living in the home, all the family participants were two-parent nuclear families. Eighteen families could be described, in terms of income and education, as middle-class, while two families could be described (and as they defined themselves during interviews) as working-class. Each family had between one and four children, ranging between ages one and twenty-two, and all adult couples were heterosexual and married. In selecting the participants, the only set of criteria that was followed came from the discipline of building engineering, and it was related to the size of the house and to the technical requirements for the monitoring equipment installation; thus, houses of similar sizes (three to four bedrooms) were chosen to make data comparison possible.

All the different disciplinary branches of the project conducted various forms of research with the twenty family participants following a pre-agreed schedule. The first research encounter was the Getting to Know You interview, which was developed by the colleagues from design. It consisted of two researchers (one from design and one from social sciences) visiting the family for one evening and joining them for a takeaway meal (paid for by the project) while carrying out a semi-structured interview. After the meal, the family would take part in an interactive activity that consisted of mapping the floor plans of their house according to their everyday activities and their daily pathways through the home. By taking part in several such encounters, I met some of the families who later became the key participants in my own study and had the opportunity to experience evening time and a meal for the first time with UK families inside their homes. Later, towards the end of my fieldwork, one family, the Smiths, invited me several times for dinner and for lunch. I also invited them to see my new place, meet my boyfriend, and try my grandma's apple cake at a later point when I was writing my dissertation.

The second research encounter was the video tour of the house (Pink 2004), which was conducted by the social sciences branch. The majority of these tours were conducted by my colleague Kerstin and two were conducted by me. Even if this research stage was part of the LEEDR overall research design and not of my doctoral research design, the knowledge gained from these encounters informed my future research questions and gave me a new perspective on the relationship between academic and applied anthropology where a focus on ethnographic encounters could open up a discussion about the similarities, rather than the differences, between academic and applied research. After every Getting to Know You and Video Tour research encounters, I wrote down detailed field notes—of a length varying between ten and twenty hand-written A5 pages—which represented the beginning of the process of thinking about family life in English middle-class contexts. Later, I took part in applying other social sciences methods, such as conducting video tour follow-up interviews and re-enactments of practices with video—focusing on laundry, cooking, digital media use, and bathroom practices—with my five key family participants.

The next stage in the LEEDR research was the installation of monitoring equipment, which recorded electricity and gas consumption as well as movement in the home. This was also the moment when the first year of my PhD was coming to an end, and I was preparing for a year of fieldwork. Thus, when I carried out my doctoral fieldwork, my participants were already well acquainted with the LEEDR project, with the other researchers, and (some of them) with me. It was in the capacity of a LEEDR researcher that I first met them, and this opened their doors to me. Because I represented a serious research project that they previously agreed to take part in, I was welcomed into the families' homes for pre-arranged visits—much more than I would have been allowed access as a new neighbour who was living in a shared student house, as my initial experience in Middleborough showed. In time, while my relationships with the key participants developed, I was regarded as "Roxana" rather than as "a LEEDR researcher." They came to know some details about me: that I grew up in Romania where winters are snowy and mountains are high; that my parents lived in France and had a Labrador; that I was a Morris dancer, which some of them even witnessed; and that I was part of a student society of gardening as well as the Transition Town local group. They sympathized with me when I had a bike accident; they tried my home-brewed beer and my cakes; and they knew about my holidays, weekend trips, and conference participations. However, our relationships developed on the basis of the initial trust offered to a researcher working as part of "a serious project." It is in the initial legitimation that the LEEDR project gave me that the greatest benefit of working as part of a project, rather than as a lone-researcher, stands for me.

My response to the double bind created by the two different types of expectations of the academic and applied-interdisciplinary domains was to design and carry out my doctoral project independently from the focus and methodologies of the LEEDR research while simultaneously conducting other research as part of the social sciences branch of the project. As a result, I did not need to keep both the academic and the applied aims in focus at once. I was able to momentarily suspend the concerns of the LEEDR project during my ethnographic encounters and focus on the exploration of other questions that were emerging from my fieldwork.

My relationships and experiences related to Morris dancing and to Transition Town activism continued to develop during and after my fieldwork. In order to distinguish between my official research and the knowledge gained serendipitously outside the realm of the project, I kept two types of diaries. One diary was dedicated to field notes that followed every encounter with the LEEDR families, which I wrote during daytime sitting at my desk. The other diary contained my thoughts about my out-of-office-hours experiences, which I wrote down while in bed late in the evening or early in the morning. I employed my native language, Romanian, for both these forms of note taking and introspection and used English only for quotes from conversations.

In order to reflect on my fieldwork while writing my dissertation, I used both sets of notebooks because the diaries provided a record of where my thinking and practical knowledge of English ways of engaging with the world were at specific moments when I had particularly rich research encounters. For example, it was in the diary where my first ideas about anticipation emerged, in the form of a list of examples of what I called at the time "anticipation techniques." I started to be interested after that moment in the actions of anticipation related to everyday family life and to address them in my research with the LEEDR participants. Therefore, I argue that applied research cannot be separated from other forms of learning in the field. While applied research endeavours ask a specific set of questions, the responses to these questions and the analytical work through which these responses are considered to extend existing knowledge are informed by other forms of learning that the ethnographer engages with, in the larger cultural context that surrounds and pervades the locus of the applied research project.

SENSORY ETHNOGRAPHY AS A ROUTE TO TACIT KNOWLEDGE

The concept of ordinary agency, which was introduced in the previous chapter, refers to forms of agency employed in everyday life that are non-propositional and related to unconscious, tacit, and non-intentional acts of creating meaning. Attention to ordinary agency as a form of practical knowledge involves a process of developing methodological practices and instruments that could tackle, in Harris's (2007) words, the "tacit aspects such as body techniques, skills, the senses" (2007: 12). Moreover, one needs to explain the methodological approach to ethnographic fieldwork that would permit the tacit knowledge the researcher and participants accessed and used in the development of their ethnographic relationship to be considered a valid form of knowledge.

During my fieldwork, folk dancing proved to be an important form of training that allowed me to "skilfully see" (Grasseni 2007) and develop my sensitivity to the ways of engagement with the world that my participants tacitly held and enacted. I employed the methodological approach of sensory ethnography to methodologically integrate this experience in relation to my fieldwork with the LEEDR families as well as to show that, even if performed by wearing an "applied hat," my ethnographic encounters with the project participants were as rich in sensory and tacit information as the process of learning folk dances.

Sensory ethnography is an approach to ethnographic practice that accounts for the sensoriality of human experience by maintaining a double focus. It focuses on the ways in which research participants explain, represent, and categorise their multisensory everyday experiences and on the ways in which the ethnographer experiences and learns about other people's lives through his or her own sensorium (Pink 2009). Following a phenomenological tradition—especially Merleau-Ponty's phenomenology of embodiment that regards senses as being interconnected—and drawing from the works of Ingold (2000), Stoller (1989, 1997) and Seremetakis (1996), sensory ethnography responds to two important contemporary epistemological considerations in social sciences. The first looks at *what* we consider to be the knowledge, the form of "data," that we can plan to acquire from research participants. The second consideration asks *how* anthropological knowledge is produced: what ethnographic research is and how an anthropologist can arrive at an understanding of the lives of others.

My doctoral fieldwork with the LEEDR families can be regarded as being comprised of two main stages. First, I employed what can be called traditional ethnographic methods, such as semi-structured interviews and participant observation, which I reinterpreted through a sensory ethnographic approach. Based on my field notes from this first stage, I further developed a set of inventive participant-led methods: the Evening Times video diaries, a method of self-interview called Five Cups of Tea, and the Tactile Time collage (these methods will be described in the next section). There were eighteen families from a total of twenty LEEDR family participants who took part in the first stage of research.[4] At the same time, after negotiations with the project team, I was allowed to pick five families with whom to carry out long-term ethnographic fieldwork. I chose the families that I already knew from previous research encounters and who I felt I would get along well with because of shared interests. Some of the families that I would have liked to work with declined the proposition, while others volunteered of their own accord after hearing from other LEEDR researchers that I was doing a selection for my PhD. The families that I ended up working with were a mixture of choice and chance. Throughout this book, when I talk about the key participants in my research, I refer to the five families who employed the set of inventive methods during the second stage of my field work while also engaging in long-term ethnographic relationships with me.

The traditional method of participant observation refers to the ethnographer's participation in the activities that take place in the everyday lives of the members of the community that are part of the research. This method is considered in ethnographic practice to be both a way of understanding how other people enact and give meaning to their daily activities and a proof of "having been there" that will legitimate the researcher's further writings. From a sensory ethnography perspective, this participation can represent a form of sensory apprenticeship because the ethnographer learns "how to sense one's environment in a culturally specific way" (Pink 2009: 70) and because they learn new skills, such as the "skilled vision" of being able to tell good cattle from bad (Grasseni 2004).

The interview as event, from a sensory ethnographic perspective, "creates a place in which to reflect, define and communicate about experiences" (Pink 2009: 87). In my research, the event created by the interview brought about the opportunity for participant observation as well. With the two exceptions of male participants who worked in academia and preferred to be interviewed at work in their offices, all the other interviews took place inside the participants' homes, such as in the living room, kitchen, or conservatory. Pink suggests that interviews that take place inside the participants' homes can bring a special kind of engagement and understanding: "By sitting with another person in their living room, in *their* chair, drinking *their* coffee from one of *their* mugs, one begins in some small way to occupy the world in a way that is similar to them" (2009: 86, original italics).

Nearly half of the interviews that I conducted were part of wider LEEDR research visits, which meant that some of my colleagues were video recording other family members upstairs or were checking or installing monitoring equipment while I was interviewing one of the adult participants downstairs. In these cases, all family members frequently engaged with various parts of the interview and spontaneously gave their opinion on one or more questions if, for example, when coming in the room, they heard the main interviewees talking with me about a topic that they were interested in. For the rest of the interviews, I was the only visitor, and one or both of the adult participants sat down with me and responded to my questions. The interviews addressed the topic of media and digital device usage at home in relation to domestic time; the duration was between one and two hours.

These visits gave me the chance to experience and be part of the home environments of the people I was studying. Family rhythms continued, and everyday events happened in my presence, such as family members going in and out of the house; having drinks, snacks, or meals; using their mobile or landline phones, computers, and tablets; watching TV or doing homework; going in and out the living room; or checking on other family members by using mobile phones or by going to the rooms that they were in. During these visits, spontaneous questions emerged from some particular features of their homes, from family photos or from occasional displays of cards. We had cups of tea together, checked the weather on mobile digital devices, laughed and wondered about what smartphones can do nowadays. Sitting with them, I experienced the view from their windows, the forms of lighting in the house, and the familiar sounds of the central heating going off, the doorbell, or the landline phone. During these visits, I felt that I was momentarily part of the worlds of the people who took part in my research. I also realized that every home was different. The sixteen houses that I visited gave me a small experiential repertoire of English homes, which I broadened by using various hospitability networks for travelling in different parts of the UK. Spending time with people inside their houses could thus be regarded as a form of training for the ability to skilfully see the types of relationships that people have with the environment of their homes, with their domestic spaces and domestic time.

While the sensory and tacit knowledge that I accessed through these encounters functioned as a sort of experiential repertoire, as a grounding layer that later allowed me to make connections when I saw similar forms of domestic engagements and relations, the opportunity to develop long-term relationships with the key family participants brought me to agency: by making me feel like an agent, the mutuality of and insight gained from these engagements opened up the idea of ordinary agency.

INVENTIVE METHODS AND OTHER OCCASIONS FOR ETHNOGRAPHIC ENCOUNTERS

After I conducted the initial set of semi-structured interviews, the need to develop other methods came both from the non-propositional particularity of the subject that I was trying to tackle—domestic time—and from the expectations that the LEEDR project had raised for the research participants. The initial engagements with the LEEDR participants created the expectation for them that taking part in research is exciting because they started with the Getting to Know You visit that involved a takeaway meal and an interactive task, which the participants really enjoyed. For example, one of these visits that I took part in lasted for over four hours, not at our request but because the participants enjoyed the task. This expectation, combined with the particular sense of privacy surrounding domestic settings that I discovered in Middleborough, made me think in the first few months of my fieldwork that I needed to provide good reasons in order to visit the homes of the family participants again. I developed the method of the Tactile Time collage to create a fun experience for my key participants, which would let me inside their homes again and might raise some new questions in relation to my research topic. At this point, I was free to develop ethnographic relationships without the involvement of the LEEDR project; people knew that my visits were related to my doctoral research and could decide not to take part in the tasks that I was proposing without this decision affecting their participation in the overall project. This situation was similar to the experiences of doing traditional fieldwork, except for the fact that I was allowed only five family participants to try this type of long-term ethnographic engagement with. In the case that they would have opted out of my research, my options would have been to analyse the existent LEEDR materials for my dissertation or change my topic to a field that I had succeeded in developing relations in, such as local environmental activism or folk dancing.

The Tactile Time collage was a form of focus group that took place in the participants' domestic settings, and it asked family members to make a series of collages together that showed the ways in which they spent time at home around various digital devices. For this task, I provided a kit that contained cardboard sheets; pictures of digital devices, household objects, and food and beverage items; coloured crayons; and textile fabrics. I introduced the collage as a representation of a clock that placed a main digital device in the centre, such as the living room TV set. I asked people to write down the times of day when they used this device and to illustrate using photos the types of activities that they performed around that device, such as using a smartphone or having a snack while watching TV. When this stage was completed, the participants were asked to express the qualities of the different moments that they had identified by choosing one or several textile fabrics from a selection of nine different textures, from hopsack to cotton and felt. They enjoyed this stage, taking their time to feel all the available textile fabrics and choose one or more that best expressed the way some moments usually felt to them. I video recorded this stage and asked them to explain their choice of fabric.

For the research participants, the collage-making activity gave them the opportunity to reflect over family routines and to talk to the other family members about the ways in which they experienced time at different moments of the day. For example, Elaine said about choosing a piece of felt to illustrate her lunchtime by the TV: "I like this one because it feels smooth and calm. I'll be on my own eating my lunch and that's how I'll be feeling: nice and calm." Meanwhile, her daughter Becky explained her choice of hopsack to illustrate TV watching after coming from school thusly: "Hopsack because I had a hard day at school and I'm feeling rough." Four families took part in the Tactile Time collage activity and made between two (illustrating TV and computer usage) and four (illustrating TV, computer, stereo system, and Wii usage) collages. In one case, it was just the female adult, Sam, who took part in the activity, making the series of collages and representing the routines of the other family members. In all the other cases, there were between three and four family members collaborating on the task, and in one case included the routines of a family member who was absent during this research encounter.

By linking the intangible experience of time with the sense of touch, my research questions about time were transformed: it suddenly became possible to address them. I found myself having a conversation with the key participants about domestic time; however, it was not a conversation the involved words, rather, it involved sensory information. As well as tacit knowledge, this form of conversation involved our capacities for imagination and articulated a different entry point into the topic of domestic time than the previous interviews that I had conducted. The Tactile Time collage revealed that domestic time is made of different types of time, each with different qualities that are sometimes intersubjective and sometimes individual. On some occasions, shared time was represented as being experienced in the same way, for example, by using a piece of felt fabric to suggest cosines and warmth. On other occasions, participants represented their experience of shared time differently than other family members did. The collage also showed that one's approach to domestic time differs with the presence or absence of other family members. In this way, it opened up new questions that I was able to explore by employing additional methods.

The next method that I proposed to and developed with the people who took part in the research addressed shared evening and weekend family time. This was a form of video diary called the Evening Times self-video recording. It was a participant-led method that asked family members to record short video clips of their communal activities around various digital media. I provided families with a small video camera (a Sony Bloggie) and asked them to share it between family members so that each person would film the others at least once over a period of a week. All five key family participants took part in this task; total recording times were between five and ten minutes, except for one family who produced videos of a total length of 52 minutes.

The resulting video clips show how family time looks like when no one outside the family is present. By capturing and showing the spontaneity of domestic life, they produce a different form of knowledge than my interviews that, for example, asked people to think about and formulate a general response that could account for what happens daily or "normally." The clips also show the responses of family members to each other's action of filming, which are revealing of the nature of the interactions and relationships between them in a more intimate way than what can be learned through interviews and participant observation. Watching the self-video recordings was also utterly surprising for me in terms of their heterogeneity. The recordings are so different, in both content and aesthetics, from one family to another that I found it difficult to think about analysing them as a set of materials that could embody communal themes. They are rather like different trips into different realities, each with its own colors, lighting, form of movement, soundtrack, types of engagements, and ways of acting. The clips show the perspectives that people had of their family time *during* this family time. By being part of family time, the act of video recording the other family members was framed by a particular set of emotions, dynamics, individual and collective habits, intersubjective meanings, and expectations, which was particular to this form, and only this form, of sociality and experience that can be called family time. The video recordings are not just representations but products of family time. The video camera that I gave to the participants became just one of the multiple digital devices that they were able to use in the evenings and on weekends. It was, therefore, integrated into their routines much more smoothly than I had imagined it would be.

The Evening Times self-video recording activity created the opportunity for two visits: one for bringing the camera and one for picking it up. One family was not happy with the quality of the resulted materials and chose to repeat the exercise using their own video camera. This provided the opportunity for me and Elaine to meet in a cafe in the town centre a few times to exchange the materials, which were on a USB stick, over a cup of coffee and a chat.

In the meantime, while we were getting to know one another better, other occasions to visit the key participants came up, such as for bringing them souvenirs from my trips to Romania and France, or for conducting other research for LEEDR, such as filming the performance and re-enactment of a set of domestic practices (cooking, doing the laundry, using digital media, and bathroom practices). Some participants watched me dancing with my Morris side in the town centre or in the town's main park, and Elaine's husband Chris regularly attended the meetings of Transition Middleborough and of a progressive political discussions group that I also attended.

The next research encounter I proposed as part of my doctoral fieldwork was an interview on childhood memories about using media at home, such as watching TV, listening to the radio, or buying, recording, and listening to music albums. This interview was a way for me to know the participants better, and I later looked for and listened to their teenage music preferences as a way of aurally accessing something from their experiences of buying and listening to their first album as youngsters. These interviews also created a space for self-reflection and for developing comparisons between the ways in which they experienced family life as children and the ways in which they experienced it now as adults. In other words, they developed comparisons between the families that they grew up in, which they did not choose and where they did not necessarily have the agency to change the things that they did not agree with, and the families that they made now. The memory interview took place towards the end of my fieldwork when the key-participants and I knew each other better.

The last research method developed as part of my ethnographic field-work was the Five Cups of Tea activity of self-interviewing with video. This method involved using a small video camera—the Sony Bloggie that the research participants had previously used for the Evening Times activity—during the preparation and drinking of a cup of tea. There was a set of two questions to respond to each time: one about the activities that they did before having the cup of tea and planned to do after having the cup of tea, and one that asked them to describe how they made their cup of tea while they made it. The third question part of the Five Cups of Tea activity was different with every cup and was responded to while sitting down and having the drink. This participant-led method looked at domestic daytime and moments of solitude, and it only addressed the female adult participants.

I reached the idea of developing the Five Cups of Tea research method when, looking back at my field notes and the materials that resulted from my fieldwork, I arrived at the topic of mothering. It seemed that some important motivations, explanations, and, contradictions in the ways female participants talked about their domestic activities were related to the individual and cultural meanings attributed to being a mother. I also discovered that most of my materials were about family activities and uniquely represented domesticity as a locus for family. These materials provided very few insights about how the domestic environment was experienced in solitude when the other family members were not at home. This was because many research encounters involved several family members who, even if not actively participating, were still making the home a collective social space and because female participants sometimes talked more about the other family members than about themselves during one-to-one encounters.

The Five Cups of Tea task created a situation of self-interviewing that was flexible and non-intrusive. It explored the moments of reflexivity brought out by the ordinary event of having a cup of tea, an event that marked a break in activities and produced a wished-for change in the participants' bodily and emotional state. I wanted to create a moment with this method when participants could distance themselves from the constant requirements and expectations related to having a family-style lifestyle and from the everyday chores that were embedded in their domestic lives. The five sets of questions participants responded to while having a cup of tea were not about domestic routines, but about personal experiences and memories. With the first two sets of questions, I intended to bring mindfulness and self-reflexivity to the experience of having a cup of tea. I asked the participants to describe the taste of the tea and if they remembered their first cup of tea and the way in which their parents used to prepare tea. The next three sets of questions were more personal and unusual to ask in a face-to-face conversation. First, I asked them to tell the camera something that they wished for in that moment. Then, I asked what kind of mothers they were and if they could describe one thing that they were sure that they were doing well and one thing they thought they were not doing well enough as mothers. Finally, I asked how they would like the world to change.

One could argue that these questions, or specifically the questions about being a mother, invited practices of problematization, or ethical work (Foucault 1990). They do indeed, but the particular way in which they were responded to, through the video camera, showed that problematization is more than a work of conscious self-reflection because it also engages the sensory and emotional experiences related to having a cup of tea and to sitting down in solitude. The video camera is a witness that shows the multisensory and emotional dimensions of ethical work in the moment that it takes place and in the context of completing the self-interviewing task that represents its trigger. The recordings that people made during the Five Cups of Tea task show problematization in Moore's definition, as containing multiple engagements with imagination, the material world, and tacit knowledge (2011: 21).

In defining the methods described in this section as inventive, I followed the conceptualisation of inventive methods sketched by Lury and Wakeford (2012) as methods engaging the social world and not investigating it. They argue that some research methods, such as the examples discussed in their edited collection (the anecdote, the experiment, probes, and speculation, among others) can be considered inventive if what emerges from their application can change the problem that the method addresses. The inventiveness, therefore, is not inherent to the method and cannot be known in advance of its use, but it can sometimes emerge from the method's application. For Lury and Wakeford, inventive methods can produce subversive disruptions, because they "have the capacity to display a kind of self-displacing movement, that is, they comprise processes of imitation and repetition in which a surplus is created that allows an event—what happens, the happening of social life—to become inventive" (2012: 7).

This depiction of inventive methods probably best suits the Five Cups of Tea activity, the enactment of which, even if based on the ordinary event of having a cup of tea, redefined this event as a moment for self-reflection and gave voice to thoughts that might have previously accompanied this type of event in a latent, interior, or tacit form. The application of the Evening Times self-video recording activity also generated inventiveness. It created new situations during family time, such as that of "mum as a filmmaker," which produced unique responses and reactions that could not have been predicted before the method was used. In developing the Tactile Time collage method, however, I was particularly inspired by the ways in which the British artist Richard Hamilton used the technique of collage to raise questions about domesticity, female identity, and consumerism in the late 1950s and early 1960s. Hamilton's use of collage was intentionally subversive because he brought together representations of familiar, everyday domestic objects in unexpected compositions that have the power to displace and trigger the questioning of a taken-for-granted domesticity, for example, in the work $he (1958–1961). While intentionally inventive from the start, as an arts-based methodological proposition, I argue that the application of the Tactile Time collage method also produced inventiveness in its performance, particularly at the textile-selection stage. During this stage, the people taking part in the research gradually learned that they could express the qualities of time through the tactility of textile fabrics, while I discovered that a conversation about domestic time has just become possible through tactile mediation.

The knowledge that was produced by employing these methods as part of my ethnographic fieldwork is discussed in the following chapters.

NOTES

1. The name of the town where my field site was located is a pseudonym.
2. See Barry (2005) and the visual narrative of the Opening Ceremony show for the 2012 Olympic Games in London.
3. In the meantime, I learned that words referring to body parts, such as "arms" and "head," are indicative of British pub names, and I met enough druids to believe that they might possibly be the nicest and warmest group of people in the UK and, indeed, might often gather in pubs of the same name.
4. While all the families were invited to take part in my research, two families, both with very young children, declined.

REFERENCES

Barry, Desmond. 2005. "A Job on the Line." *Granta (89): The Factory.*
Bloch, Maurice. 1985. "From Cognition to Ideology." In *Power and Knowledge: Anthropological and Sociological Approaches*, edited by Richard Fardon, 21–48. Edinburgh: Scottish Academic Press.
Borofsky, R. 2000. "To Laugh or to Cry?" *Anthropology News* February: 9–10.
Drazin, Adam. 2006. "The Need to Engage with Non-Ethnographic Research Methods: A Personal View." In *Applications of Anthropology: Professional Anthropology in the Twenty-First Century*, edited by Sarah Pink. Oxford: Berghahn Books.
Edwards, Jeanette. 2000. *Born and Bred: Idioms of Kinship and New Reproductive Technologies in England.* Oxford: Oxford University Press.
Foucault, Michel. 1990. *The History of Sexuality*, vol. 2. London: Penguin Books.
Grasseni, Cristina. 2004. "Skilled Visions: An Apprenticeship in Breeding Aesthetics." *Social Anthropology* 12: 41–55.
Grasseni, Cristina. 2007. "Communities of Practice and Forms of Life: Towards a Rehabilitation of Vision?" In *Ways of Knowing*, edited by Mark Harris, 201–223. Oxford: Berghahn Books.
Harris, Mark. 2007. "Introduction: 'Ways of Knowing.'" In *Ways of Knowing: New Approaches in the Anthropology of Experience and Learning*, edited by Mark Harris, 1–24. Oxford: Berghahn Books.
Ingold, Tim. 2000. *The Perception of the Environment.* London: Routledge.
Jewell, Helen. 1994. *The North-South Divide: The Origins of Northern Consciousness in England.* Manchester: Manchester University Press.
Kligman, Gail. 1981. *Calus: Symbolic Transformation in Romanian Ritual.* Chicago: University of Chicago Press.
Lewis, Jim, and Alan Townsend. 1989. "Introduction." In *The North-South Divide: Regional Change in Britain in the 1980s*, edited by Jim Lewis and Alan Townsend, 1–20. London: Paul Chapman Publishing.

Lury, Celia, and Nina Wakeford. 2012. "Introduction: A Perpetual Inventory." In *Inventive Methods: The Happening of the Social*, edited by Celia Lury and Nina Wakeford, 1–24. London: Routledge.

Moore, Henrietta L. 2011. *Still Life: Hopes, Desires and Satisfactions*. Cambridge: Polity Press.

Pink, Sarah. 2004. *Home Truths: Gender, Domestic Objects and Everyday Life*. Oxford: Berg.

Pink, Sarah. 2009. *Doing Sensory Ethnography*. London: Sage.

Rapport, Nigel. 1993. *Diverse World-Views in an English Village*. Edinburgh: Edinburgh University Press.

Roberts, S. 2006. "The Pure and the Impure? Reflections on Applying Anthropology and Doing Ethnography." In *Applications of Anthropology: Professional Anthropology in the Twenty-First Century*, edited by Sarah Pink. Oxford: Berghahn Books.

Ryle, Gilbert. 1949. *The Concept of Mind*. Chicago: University of Chicago Press.

Seremetakis, N. 1996. "The Memory of the Senses: Historical Perception, Commensal Exchange, and Modernity'." In *Visualizing Theory*, edited by L. Taylor. London: Routledge.

Singer, M. 2000. "Why I Am Not a Public Anthropologist." *Anthropology News* 6–7, September.

Stoller, P. 1989. *The Taste of Ethnographic Things: The Senses in Ethnography*. Philadelphia, PA: University of Philadelphia Press.

Stoller, P. 1997. *Sensuous Scholarship*. Philadelphia, PA: University of Philadelphia Press.

Strathern, Marilyn. 1981. *Kinship at the Core: An Anthropology of Elmdon, a Village in the N-W Essex in the 1960s*. Cambridge: Cambridge University Press.

Sunderland, P.L., and R.M. Denny. 2007. *Doing Anthropology in Consumer Research*. Walnut Creek, CA: Left Coast Press.

Williams, Raymond. 1973. *The Country and the City*. New York: Oxford University Press.

Wright, S. 2006. "Machetes into a Jungle? A History of Anthropology in Policy and Practice, 1981–2000." In *Applications of Anthropology: Professional Anthropology in the Twenty-First Century*, edited by Sarah Pink. Oxford: Berghahn Books.

CHAPTER 4

Meeting the Families

The Smiths, the Nicholls, Joyce and Erin Love, the Hewitt-Mitchells, and the Johnsons were the five key family participants who engaged in my overall research and that I developed additional "inventive" methods with. Before continuing with the empirical chapters that discuss spontaneity, anticipation and, "family time," respectively, I would like to introduce them. To do so I will make use of, besides my own writing, a quote from each family extracted from the Five Cups of Tea activity. These extracts are the answers that the female adult participants gave to the questions, "What kind of mum are you or do you try to be? Could you please tell me one thing that you are sure you are good at and one thing that you are not sure you are doing well as a mum?" The Five Cups of Tea activity was a self-interviewing method using video, and the participants answered this set of questions, among others, in a moment of solitude when they were at home having a cup of tea. While the question specifically addresses the topic of mothering, the answers refer to general family dynamics as well, and this is one of the reasons for including them here.

The video recordings that emerged from the Five Cups of Tea method were made towards the end of my fieldwork. Watching them proved to be an important experience for the ways in which I subsequently ethnographically described and interpreted my research. The recordings revealed a form of epistolary intimacy in the relationship between the ethnographer and the participants, manifested when the physically absent researcher

© The Editor(s) (if applicable) and The Author(s) 2016   73
R. Moroşanu, *An Ethnography of Household Energy Demand in the UK*, DOI 10.1057/978-1-137-59341-2_4

was addressed through the medium that she provided: a video camera. Together with the propositional content of the answers, the multisensory and affective elements that are communicated in the recordings—such as the setting and lighting, the tone of voice, and the emotion in one's voice—made me feel that I would be able to imagine some aspects of the ways in which these people, who kindly and generously participated in my research, regarded and experienced their lives. By including these extracts here, therefore, I wish to express the wonder and affection I felt during my fieldwork while also reflecting on the unique potential for empowerment that ethnographic relationships have.

LA MAISON DU BONHEUR

Now in their late forties, Cynthia and Jeff Smith moved to Middleborough when their oldest daughter Kelly was a toddler. Jeff got a job as a university teacher in engineering at the local university while Cynthia taught French and German at a secondary school. They got a mortgage for a detached house in a residential area located on the forest side of the town near the university's campus. They had two more children, Lee and Emma, while living in Middleborough. When I first met them in September 2011, Kelly was nineteen, Lee was seventeen, and Emma was ten. However, during and after my fieldwork (my ethnographic relationship with the Smiths developed into a friendship that continued for the entire time I lived in the town) I witnessed Lee successfully taking his A-levels and going to university, Emma starting secondary school, and Kelly moving out to find a permanent job and getting engaged.

Cynthia and Jeff first met when they were students in Manchester and members of a student sailing society. They both enjoyed sailing, but only Jeff kept this hobby after moving to Middleborough while Cynthia discovered other activities. She joined an all-female Border Morris side and later got involved in a Brownies club that her daughters were attending. Realizing that he did not have the necessary time to fully participate in and enjoy the sailing club, which was located about 30 miles from the town, Jeff later decided to give up his membership and started attending a local photography club instead. He became very passionate about photography, and his photos won several competitions. They were exhibited in the local Town Hall as well as other venues in the county. It was easy for the family to find the right Christmas and birthday presents for Jeff because additional photography equipment and accessories were always needed. Cynthia was an avid reader, and because her family bought her a Kindle as a Christmas present, she was happy with being gifted Amazon vouchers for any occasion that she would use to buy new books. She was the co-author of two language-learning books for children. Kelly liked working with children and she volunteered with the Brownies group that her younger sister attended, helping with organizing and joining them for camping trips and other activities. Lee was involved in a fan-fiction website, where a community of readers of a popular youth book series were rewriting and proposing continuations of the storyline. Emma attended a dancing club and was awarded a prize for the best dancer in her age group one year. She also played the French horn. I once joined Cynthia at one of Emma's performances in the musical *Joseph and the Amazing Technicolor Dream-Coat*, which took place in the hall of an arts college in a nearby village. In the summer before starting secondary school, Cynthia and Jeff let Emma choose a couple of pet rabbits from a rescue centre. She was very excited about her new fluffy pets because they only had tropical fish as pets before. She was so proud to show them to me to the point that we once spent an entire afternoon in their garden watching and playing with them.

The Smiths had a campervan and they were very fond of going away on big family holidays. They said they would always choose to spend their savings on a family holiday rather than on projects to make their house perfect, such as redecorating or building an extension. For Jeff, this was a conscious decision that came in opposition to what his own parents valued. He recalled that they never went away when he was growing up because his parents preferred to spend their money and their time on improving the house. Cynthia and Jeff often took their children camping and went on a longer trip once a year that sometimes involved crossing the English Channel on a ferry. Family holidays for them were a time of bonding, having new shared experiences, and creating memories. Holidays were a very special form of family time that they valued much more than a set of material goods.

Both Jeff and Cynthia were good examples of social mobility. Their extended family was proud of the fact that they went to university and of their jobs in teaching—they believed that hard work could bring success. They strongly encouraged their children to study because they knew that this could make a big difference for their futures. They were surprised, therefore, when Kelly rebelled, moved out with her boyfriend, and got a job in a supermarket at eighteen years old. However, when I first met them, Kelly was living back with the family while deciding, with the support of her parents, about what she wanted to do and what type of qualification she was going to pursue.

Concurrent with Kelly's decision to move out, Cynthia had to change her job. She went through a retraining programme and received a qualification in primary school teaching. This took time, challenged her, lowered the household's income for a while, and was generally not the best context for their oldest child to fly the nest frantically. Cynthia's first job in primary school teaching was a supplementary position. In her words, this involved a lot of waiting for the phone call that would let her know if she was needed that day. This is when I first met her, during this period of transition when she was spending much more time at home waiting than she would have liked. This meant that my research visits, with funky methods and questions, were bringing a change into Cynthia's day, between checking job websites and networking with her former classmates from the requalification course. Soon after that she got a part-time job, and she embarked upon a full-time permanent position the following school year.

After I finished my fieldwork, the Smiths and I continued to visit each other because I continued living in Middleborough. Once, when Cynthia, Jeff, and Emma visited us, my partner brought down from the attic his own French horn that he used to play when he was in school. Emma tried it and spent the rest of the visit surrounded by music scores, choosing the ones she wanted to take home with her. It was different to see them as a family of three—with Kelly and Lee having moved away—after knowing about the busyness and liveliness of their home of five. A small metal plate that I brought them from a holiday in France said "La maison du bonheur". They were still the house described there, but happiness was now quieter and calmer.

Later, after I finished my PhD and started applying for jobs, I realized how lucky I was to have met Cynthia during what for her were the uncertain times of changing jobs and see her succeeding. Her determination inspired me to approach the process of job hunting one day at a time and with optimism. "I had ten job interviews before I found this job, so you need to keep on going," she told me when we met at a feedback event that marked the end on the LEEDR project.

> I think I'm a very practical sort of mum. I think it's a job; I think it's my job to make sure that my children are ready for the big wide world out there. I have very high expectations of them, which leads to the thing I'm not sure I'm doing very well. I have very high expectations of my children,

which sometimes means I am a bit critical of them. I have to remember to be more positive. I'm sure I'm good at being there for them and listening. Not sure I'm very good at the teenage thing. Very good when they're little and I think I'm pretty good once they've grown up. It's the little bit in the middle, which tends to get a bit negative. So, yeah, generally I'm a bit of a tiger mum; I defend my children for everything against anybody.

ALLOTMENT PRODUCE

The Nicholls consisted of Sam, her husband Peter, their children Alex and Julie, and the recent addition of Ollie, a puddle-cooker spaniel crossbreed. They lived in the same neighbourhood with the Smiths; Julie was Emma's age, and they went to primary school together while Alex was two years younger. Sam turned 40 during the LEEDR research, and Peter was in his mid-forties.

Sam and Peter liked sports. They were both big football fans, and one of their favourite weekend activities was going together with their kids to the matches of the main football team in the county. Sam had a karate black belt, went to the gym, and cycled to work nearly every day. For Peter, keeping fit, which he did by jogging around their neighbourhood, was also important for his job in the police force.

When she was eight, Sam and her family moved from the south of England to a nearby village. She and Peter met in Middleborough in their early twenties, got married, and bought a small semi-detached house. They lived there for seven years until they had Julie and decided to move to a bigger house. That was when they got a mortgage for their four-bedroom detached house, which was located in the proximity of the campus. Sam worked part-time for the local university in an administrative position.

Sam and Peter were avid gardeners. Besides their back garden, where they grew herbs and vegetables and raised chickens, they also had an allotment. From March to October, they spent their Friday mornings at the allotment, digging, planting, and growing vegetables while listening to music on their portable radio or doing the Radio 2 quiz. Sam did not go to work on Fridays, so she spent the afternoon making soup, chutney, and jam from their allotment produce.

During one of my research visits, in early February, the window sill of their conservatory was covered with potatoes growing shoots and transforming into seed tubers, which Peter and Sam were planning to plant at their allotment. Another time, Peter took my researcher colleague and me upstairs to proudly show us his new seed germinator that was put to work in the spare bedroom. We often talked about gardening because I was part of a students' gardening society and was eager to learn more. During one of my visits, I asked them how they decided to take up the allotment. Peter told me that, after being on a waiting list for over a year, they got their allotment about two years before. Their main reason for taking up the plot of land was their children. He wanted to teach them how to grow their own food because he thought this would be a good skill for the future—good for them and also good for the planet. While they were still young and not particularly interested in gardening at that moment, he hoped that in time he could teach them more about how to grow their own food.

The Nicholls were one of the first families in their neighbourhood to install solar panels. They liked the idea of being self-sufficient, which they applied as much as they could to their food as well. They tried to limit the amount of products they bought from supermarkets, counting on their own fruit and vegetable produce and the eggs from their chickens. When they recently realized that they needed more eggs, they bought a third chicken, preferring this option to buying extra eggs from a supermarket. Their milk did not come from a shop either because it was delivered by a milkman.

Peter's job required shift work, which divided their domestic time into two sets of routines. When Peter was at home in the evenings, the family had dinner together around the dinner table listening to a music CD chosen by Peter and Julie. When he worked an evening shift, Sam and the children would have their dinner in front of the TV. As the shift work requirement put a strain on their everyday family time, Sam and Peter tried to make up for this limitation by organizing numerous family holidays. They always went away for the children's half-term week long breaks, camping in the UK or travelling abroad. During their participation in the LEEDR project, they also organized two big family holidays that they approached as important events and talked about a lot in anticipation as well as afterwards. The first big holiday was for visiting the theme parks in Florida where one of the highlights was swimming with dolphins. The second one, two years later, was a cruise on the Mediterranean with many stops for sightseeing.

Sam and Peter encouraged their children to be as active and sporty as they were, and the acquisition of Ollie was a step in this direction too; it was meant to make the kids spend more time playing outside. When I first met them, Julie was attending a drama club and was very passionate about acting and singing. Together with a couple of friends, I went to see her performing in the Wizard of Oz at the local Town Hall. Later, she started making jewellery, which she sold through her website with the help of her mum and at car boot sales where Sam also sold her jam and chutney. I bought a few pairs of earrings from her to gift for Christmas to my friends in Romania. Recently, when I paid them a quick visit, she told me she was planning to focus more on sport activities from now on. Alex took swimming lessons, and his favourite weekend activity was snowtubing at a nearby dry ski slope centre.

I try to be a good mum. I think I am pretty strict, probably. The kids would tell you quite strict—I do shout a lot. But I am trying to do what's best for them and try to be as understanding, let them do as much as they can on their own. I am very good at keeping them safe. Julie tells me I'm good at cooking. I try to be understanding, I try to be... I don't know, it's quite a hard thing to say... I try to be as better mum as I can. I try to learn from what I felt when I was a teenager towards my parents; I try to either copy what that they did, or if it was something that I thought, "Oh, why did my mum let me go out?" or something like that, than I try not to do that. So, I try to learn from my childhood and pass it on to them. Although that it's hard now, with the digital age, they have so much TV etcetera to sit in front of. I try to be there for them all the time. I suppose that's one thing I'm good at: is that I'm always pretty much there because I'm here when they get home from school, and I see them off to school. So there's always someone at home, pretty much, most of the time. What I'm not doing so well... Huh, I'm not keeping my patience. I do get annoyed very easily and I end up shouting, mainly because the kids aren't listening when I ask them to do something. So, I suppose that's my impatience, because if I say, "Can you do something?" I really do expect it to be done straight away rather than asking four or five times. So, I suppose I need to try to get better at being patient and stop the shouting. But I don't do it too bad; you'll have to ask the kids.

SIOUXSIE AND THE HULA HOOPS

The kitchen was bright and quiet. Towels and bed sheets were drying on a couple of laundry racks around the kitchen table. The green of the garden, through the large windows, surrounded the room like wrapping paper. It was June and a Saturday morning. At the table, Joyce used her laptop to video chat with Erin.

Her daughter chose a picture of pink petals to frame her video image. Joyce was looking for a frame, showing me the options to be found in the settings of the video software. She picked a frame of seashells. "It's very sunny in here," Erin said, and her video image turned brighter. "Now it's rainy," she said, making the image darker. Joyce was one step behind, not sure where the settings for brightness were. She found them and showed me how to choose between different weather conditions. In the meantime, Erin turned upside down; we laughed while she lifted up her arms making it look as if she was hanging down from the ceiling. All of a sudden, it seemed that the video image of Joyce was frozen. Then Erin had to come downstairs to show her mum how to unfreeze it.

We spent over an hour in this way, with Joyce and Erin video chatting between the kitchen and the bedroom while I filmed them in turns for a LEEDR research encounter focused on digital media practices. In their video chatting, the mum and daughter were happy to just look at each other, change frames, and giggle, not exchanging many words; it felt like a relaxed time of hanging around on a Saturday morning. A pair of roller skates and a unicycle were waiting by the door. The living room table was covered in multi-coloured origami creations. A couple of hula hoops were resting by the fireplace.

Joyce and Erin Love lived by themselves in a detached house, close to the town centre. In her mid-late thirties, Joyce was a homemaker, and most of her activities were focused on Erin's upbringing.

> I try to be a supportive mum. I try to be there for me daughter. I certainly don't try to do the heavy-handed authoritarian sort of mum. I probably try to be... Well we are pretty close and I like to think that she can talk to me about pretty much anything, come to me if she had problems. So, yeah, I just try to be there, kind of in the background, doing stuff just to make life easier for her 'cause I think it's pretty stressful these days: school, activities, quite full on. One thing I'm fairly sure I'm good at doing is listening to my daughter if she has problems. I think I'm quite good at trying to think of solutions, to try to get around some of those things. Or, even better, I like to think I try to encourage my daughter to think of them. Something that I'm really not very good at would definitely be timings of things, for example bed times; not good at getting a consistent bed time, and often they tend to be way too late, which is not good.

Joyce grew up in Middleborough after having moved here with her family at the age of six when her dad got a job at the local college. She knew the town very well because she changed a few houses during different life stages. Her mother passed away when Joyce was in her early twenties. Subsequently, after dividing the family's possessions between the three of them, her father moved to London, her brother moved to America, and Joyce stayed in the town. She kept a painting made by her mum, of a busy street, in a Lowry style hanging on a wall in the landing. She often cooked her mum's recipes, most of them vegetarian. She liked and tried to cook Asian food as well, sometimes doing her shopping for spices and ingredients at South Asian shops in Middleborough or in a nearby city. When she was a teenager, she liked listening to alternative and punk music, and she often used to go to see her favourite bands live: The Stranglers, Siouxsie and the Banshees, Joy Division, Stiff Little Fingers. She did a degree in computer science and worked in software development until she had Erin.

When I first met them, Erin was ten. She loved learning circus skills, and she also attended a dance club. We soon realized that we had some friends in common—Helena and Christine were leading the circus skills club while being part of my Morris dancing side as well. We later found each other performing at the same events, each with her group. In the summer of 2012, when the Olympic Games were held in London, the town's fête, Picnic in the Park, was bigger than usual. It started with a parade around the town centre, where my group and Erin's marched together between other groups of performers, and it continued in the town's park with a full day of performances. We hung around together and watched each other's performances while Joyce took pictures.

For Joyce, it was important to keep up with the new technologies. Even if they could not necessarily afford to buy the latest models of gadgets, they both had Android smartphones, their own laptops, and Erin had a tablet that she used to play games. Knowing how to use these devices was a form of literacy for Joyce that she considered essential for today's world. She did not want to be left behind, and she was not attached to the idea of a better past where the forms of leisure and of sociality were less superficial than today. Instead, by making these devices part of their everyday life, Joyce and Erin oriented themselves towards the future.

I sometimes met Joyce on Tuesday afternoons at the local gym when Erin went to her dance club and the mum used the waiting time for a gym session. Because I moved to the town centre, I had been inviting Joyce for a cup of tea at my new place. But my address struck her. It seemed I now lived in exactly the same house that her brother used to rent over a decade ago before he moved to America. Joyce was not particularly keen to see the house again, but she kept on telling me that she would pay me a visit someday soon.

APPLE FLAPJACKS

Lara and Dominic Hewitt-Mitchell were in their forties and early fifties, respectively. They had three children from their present and previous marriages. They lived together with their youngest, Ewan, who was eleven when I first met them, and their cat and two French Spaniel dogs. In his early twenties, Dominic's son Mick lived on his own while Justine, Lara's daughter, left home for university. However, both Mick's and Justine's bedrooms were kept intact for their weekend visits. Justine always came home for university holidays, and her bedroom was just as she left it.

The family lived in a four-bedroom detached house on the forest end of the town. Their kitchen faced a wide green field surrounded by tall trees where they took their dogs to walk every day. Lara and Dominic's bedroom was an ensuite in the attic conversion; their bed was placed against a wall displaying a large map of the world. Their kitchen had a French look, with a wall painted in blackboard paint on which they wrote shopping lists and left notes. It also had an antique birchwood kitchen cabinet that displayed a set of plates and pots made and painted by Lara during her evening pottery classes. They liked having friends around for dinner or a cup of tea, and I met some of their guests a few times when I visited them for my research. In the summer, they would move their dining table to the deck in their garden and have their meals there. Dominic also used the back garden for growing vegetables and herbs.

Lara's mum lived in Spain where she moved together with Lara's dad when they retired. He unexpectedly passed away soon after, and Lara's mum preferred to stay there, even if on her own, than to return to the UK. Lara's brother lived and worked in France in the wine industry. The Hewitt-Mitchells often visited them in Spain and France, and Lara kept in touch through email, phone, and video chatting. We often talked about places to visit in France, where my parents were also living at the time, and about dogs. One of their dogs, Ozzy, had his own Facebook profile, and I worked on an epic poem about my parents' dog Rich.

When I first met them, Dominic worked from home, doing small consultancy projects while looking for a full-time job. He later got a job in Birmingham, which required him to commute daily over a more than one-hour driving distance and changed their domestic routines. Lara was working full-time in special needs education. Ewan was attending secondary school, and his favourite hobby was playing cricket.

As Dominic spent more time at home, he took part in my initial interview about digital media and domestic time, and he made all the video recordings for the Evening Times video diary as well. However, for the Tactile Time collage, he was joined by Lara and Ewan. When it was her turn to video record and self-interview for the Five Cups of Tea activity, Lara preferred to film the preparation of the cup of tea in silence and write down the answers to the questions rather than talking to the camera.

> I think I am a good mum; I try to be "firm but fair." I am good at talking with my children; this enables them to know I am always there if they need to discuss anything. Perhaps I could help Ewan more with his homework...

Lara often cooked separately for Ewan, as he did not like some of the ingredients or dishes that she and Dominic had. She did not think about it as an extra effort and preferred letting everybody have their favourite dish rather than trying to find a dish or two that would be unanimously accepted. Once, when I dropped by to collect the video camera and the recordings they made for a research task, Lara had just finished making apple flapjacks only to find out that Ewan did not like them. "He likes flapjacks and he likes apples, but apparently he does not like apple flapjacks," she shrugged while inviting me to try the cake. "I just didn't know that, sorry!" she said to Ewan, who was playing a video game in the living room. She did not insist on the effort or the resources involved, and it looked like she did not take his refusal personally but was just surprised to find out something new about her son.

ACTION FOR A BETTER WORLD

Elaine and Chris Johnson were in their early forties. They lived together with their two children Becky and Tom, who were aged eleven and nine, respectively, and with their two cats. They were from the north of England and they first met at university in York. Elaine studied design, and Chris did a degree in engineering.

They lived in a recently developed residential area of Middleborough, which was situated further from both the town centre and the university's campus than any other LEEDR household, but was closer to one of the main A roads. They moved to Middleborough because it was a good in-between location to travel from, both towards the north and towards London, which was important for their jobs in sales. They also found a newly built detached house in a quiet neighbourhood that they liked that they could get a mortgage for. Both children's schools were within a five-minute distance from their house, and the area felt calm and safe. When visiting them, I always had to cycle because it was too difficult and time-consuming to try to reach their house by walking. They all had bikes as well and sometimes cycled as a form of exercise.

When I talked to Chris for the first time on the phone to arrange a visit after previously having emailed them, he asked me if I was part of Transition Town Middleborough because he had seen my name in the email discussion group of the organization. He was also a member of Transition and of another progressive discussion group that I used to attend at the time. At the end of my first visit to their place, he showed me an alternative cycling route, and we cycled together to the university campus, talking about corporate greed, the causes of the economic crisis, and he explained to me the differences between a bank and a building society. Chris was one of the founding members of a residents' association whose activism against planning permission for a new development that would cover all the green space that was left between their neighbourhood and the motorway appeared in the local paper several times. He was interested in hacking as a way of making things, such as solar panels, and he had subscriptions to the magazines *Make* and *Popular Mechanics* that specialized in this type of information. It would be fair to say that his interest in this form of craftsmanship came from having grown up watching his dad, who worked for a TV repairing company in the daytime and was a self-taught amateur cinematographer in his spare time, film and manually editing his cine-films in their front room. This was not just a hobby for passing the time. Even if he did not earn an income from making films, his dad won many prizes, and one of his films was selected for and screened at the Cannes Film Festival.

At the LEEDR feedback event, which took place at the end of the project in November 2014, while a set of ideas for design interventions to reduce energy consumption such as smartphone apps were being presented, Chris reacted. He pointed out that it seemed all these ideas started with the assumption that people need to be tricked to consume less energy, while the wider political context related to energy production and distribution, which was a much more important problem in his opinion, was not being considered. The speakers were confused for a moment and did not have a response at hand. But Chris was not necessarily expecting a response as such; he wanted us to look at the relationship between academia and the government, which was mediated by public funding bodies that formulated research questions and topics and subsequently funded scholars to address. When Chris pointed out this underlying issue, the entire discursive bubble that the event created by focusing on tailored feedback that would help each household reduce their energy consumption and bills was in danger of breaking. It felt like everybody in the room understood and agreed with Chris's point and were all aware that, even if we lowered our energy demand and we changed our lifestyles as much as we could, some major politico-economic transformations would still be necessary to slow climate change. In order to move on, the speakers asked the other LEEDR team members to give the participants the personalized feedback books that we prepared, and we spent the rest of the event chatting in small groups.

To say that Chris intentionally asks, in a northern accent, uncomfortable questions, is not a stereotype. Rather, it acknowledges his upbringing in a tradition of self-taught craftsmanship as well as a long (regional) history of speaking frankly and standing up for one's rights. What was surprising for me, coming from Romania where everyday talk usually includes complaining about, mocking, or criticizing individual politicians or the government as a whole, was that I did not hear a reaction like Chris's more often in Middleborough.

In her own way, Elaine followed the same ethos of frank speaking. In our research encounters, as well as in subsequent meetings in a coffee shop in town, she talked directly and with generosity about various aspects of her life. Her video recordings for the Five Cups of Tea activity were particularly illuminating and inspired me to make an ethnographic documentary film, *Mum's Cup*, based on the material that four key participants, including Elaine, filmed in their homes in response to this research method.

> So, the kind of mum I try to be, or I am… I always try to be there for them, so they have a really stable childhood, and it's always someone there that they could trust to be there for them, to tell me about all the fantastic things that are happening in their lives. And also if they need some help, I'd like to think that I'm always there for them. Because dad, he works away quite a lot. And I know that when Becky in particular was younger, and I used to

work, and dad was away a lot with his job–I mean not all the time, but he used to go away quite often through the week and be back at the weekends anyway. I don't know, it got really difficult at one point. It was just when she started school; and she got very anxious—the teachers told me about this. I don't know, there were times when I had to drop her off to friends to take her to school and things like that, 'cause I had to go to London with work and dad was away. She didn't feel very settled. And Tom was pre-school at this stage, so I was dropping him at places; it was all a bit frantic, trying to fit everything in. So, I decided to leave work because of that reason mainly; 'cause what I was doing, things had changed at the company anyway... And I just thought, "Is this all really worth it?" Because we'd had two children and we wanted to be a proper family; and dad had a good job, so I decided to take the full-time mum role. So, I would say that that's what I do; I like to be there for them. I mean, I have worked, on and off, quite often. But at the moment I'm not working 'cause the work isn't there to be honest. But I must admit I really enjoy being a full-time mum and being there for them, trying to help them. Because when you go to work, you are tired, and you have to do your chores, and often dad's away as well; and it's really hard just to fit everything in. And the things that suffer tend to be the important things, like listening to your children, helping them with their homework; little things like that, but they are quite important. So I think I'm quite good at that, actually; when I am at home, I do put all my effort into being a good role model shall we say. And both Chris and I believe in sort of traditional family values and having a really strong family unit. One thing I'm not sure I do well as a mum... I think I would say I was probably not so good at cooking. I can cook, don't get me wrong; and what I do, I do well. But I always feel that I could do more. When it comes to puddings, I never do. I always buy them, I don't make them myself; I make the odd apple crumble and things like that. But I feel like I'm not creative enough. And there are lots of creative food you can make that is child-friendly. Because I have done a lot of cooking in the past; when Chris and I were just a couple, we used to enjoy cooking in the kitchen, but obviously we don't make those kind of things now when we've got a family. So, I do tend to buy a lot of, sort of chicken goujons, and we have chips and this and that. And, I don't know, I feel as though I ought to be doing more; more proper from scratch cooking that's actually good and healthy and that the children would love and eat. I do do things, don't get me wrong, but I feel like I could do more. So, I think that's what I'm lacking in, and I just need to... Especially now that they are growing up, I need to experiment more and do some fantastic recipes.

CHAPTER 5

Spontaneity

Holding a hand-made replica of an Olympic torch, purple tissue paper ribbons as flames, I waited for my turn. My fellow dancers just started performing the Sports Dance, a dance that we created to acknowledge the local university that was renowned for their sports training. We were at Picnic in the Park, the town's fete, which takes place every year in early June. It was 2012, the year when the Olympic Games were held in London and the real torch was expected to come to our town in a few weeks' time. I hid the paper torch behind my back, inconspicuously holding it upside-down, while standing by the band. After the next chorus, the dancers would do the race figure, and that's when I had to appear. Shortly, two dancers would align in a start position and would race each other in slow motion to the other two dancers who would be standing 5 metres away holding their sticks as a finish line tape. When she handed me the purple torch, Christine said that I should appear suddenly and run in front of the two people racing, torch held high. This move was a new addition to the dance and was meant to be a way of greeting the national events. We didn't practice it because the paper torch had only been ready that morning. So, I didn't know exactly how it was going to go.

I watched the two dancers align and start the slow motion race. And then I ran. Not only holding but also revealing the torch to the audience like a sceptre, I ran with long springy steps at normal speed on an imaginary lane between the two racing dancers. I left them behind and turned to watch

© The Editor(s) (if applicable) and The Author(s) 2016
R. Moroşanu, *An Ethnography of Household Energy Demand in the UK*, DOI 10.1057/978-1-137-59341-2_5

their puzzlement as I broke the finish tape while they were still halfway through their slow-motion race. The audience laughed, and the dancers reached the finish and started laughing too. The musicians stopped playing and were very amused as well. It occurred to me that I just re-enacted a comedy trope common in the sketches of Benny Hill and the genre of English pantomime. What just happened was very surprising for everyone, including me. It was classified as a successful update of the dance, and my fellow dancers congratulated me. Subsequently, I got to perform the role of the torch-bearer in most of the dances that summer, and the paper torch was replaced by an inflatable one that enhanced the humorous effect.

It is hard to describe what spontaneity entails. As a series of actions that you just find yourself doing, spontaneity engenders emotions and sensations that you do not stop in order to observe and to translate into words. If you would stop to notice them, you would stop being spontaneous. It is not in the nature of spontaneous actions to be self-reflexive or to lend themselves to the endeavour of establishing relationships of causality—the endeavour of explanation.

The people who took part in my research often talked about the ways in which they were able to spontaneously and instantly check the weather on their smartphones, find the answer to a question insistently asked by their children, or translate inches into metres while in a shop choosing between two products. They were enthusiastic about their capacity to immediately find the information they needed because it could make their day better and lift their mood.

The significance of spontaneity in the everyday lives of my key participants emerged, particularly, in relation to actions that involved digital media. In addition to these situations, in this chapter I will also discuss examples of non-technological spontaneous actions (e.g., improvisation in Morris dancing or having a friendly argument during a family gathering) to move away from a technologically-deterministic approach that, in Green's (2002) words, makes ICTs into an a stereotype. Thus, I will not argue that digital devices are generating new types of human action. I argue, instead, that the people who took part in my research used digital devices to transform a wider variety of actions into "happy actions"— situations in which desire takes the form of unconstrained action (Taylor 1979). In the ways that spontaneity was articulated through digital media, existent cultural understandings of what it is to be an agent—of how one could convey the sense of being in control of one's life and of the ways in which one could express one's individuality—were reformulated.

Spontaneity is an approach to and expression of human action that is visible and intersubjective; it can be articulated through appropriate "techniques of the body" (Mauss 1979 [1935]) and through displays of emotion linked to shared emotional codes (Luhrmann 2006). Associated to specific forms of sociality (e.g., friendly arguments), which I will discuss below, spontaneity was expressed before the emergence of new digital devices through other recognizable sets of movements, tones, and gestures, such as those related to searching for the solution to the argument in encyclopaedias.

In addition, spontaneity depicts a specific type of relationship with time, in which the short-term future is domesticated by being transformed into action. As a distinct way of blending the present and future together, spontaneity makes the future present and opposes the logic of linear time as a dominant form of temporality that is expressed and maintained through digital clocks, among other devices that claim to provide objective representations of time. Spontaneity contradicts a managerial approach to time (Hochschild 1997) in which time is regarded as a resource that people consume (Shove 2009) and that needs to be efficiently organized. Instead, what spontaneity proposes is a playful and subjective approach. It operates by interrupting the uniform linearity of time to create wished for future moments and to make them present.

This chapter discusses a series of situations in which the spontaneity expressed in the research participants' relationships with digital media can be regarded as contributing to the extension of the "ethical imagination" (Moore 2011) of the people involved, both participants and researcher, and to developing new forms of sociality. First, I will show that the capacity to produce immediacy of new technologies was used to expand and diversify the category of what could be considered achievable wants. Second, I will discuss the ways in which the sense of an enlarged group of achievable wants is employed in developing specific forms of sociality and how it can contribute to broadening one's ethical imagination.

SPONTANEITY AND AGENTIC EMOTIONS

When the British drama therapist Peter Slade proposed child drama and improvisation play in education and therapy in the 1950s, he suggested that the ability to improvise "has something to do with drama, but is much more nearly a function of everyday life, where spontaneity, natural wit, courage in adversity, sympathy with sorrow or just sheer high spirits contribute to all that might be called the golden hour" (Slade 1968: vii). For Slade, spontaneity was both an ability that could be trained and a particular way of engagement with the world. Writing about the value of spontaneity as a capacity, Slade outlines that "part from developing the ability to speak (and incidentally to write with more imagination), other qualities become evident—a growing absorption in the task and a sincerity about the way of doing it, particularly in children, useful for all learning and general attitudes to life; also a mounting confidence and ultimately a mounting happiness" (Slade 1968: 4).

This chapter reflects Slade's view on spontaneity, specifically emphasizing two elements of his definition. The first element is the sincerity about the way of doing the task, which is an important characteristic of spontaneity that I discuss in the next section in relation to Anscombe's (1981) reinterpretation of the Aristotelian concept of "practical truth." The second element is the suggestion that spontaneity generates mounting confidence and happiness and is addressed here by looking at the forms of ordinary agency engendered by spontaneity, especially at the emotions associated with the accomplishment of "happy actions" (Taylor 1979).

These emotions are expressed below in two distinct quotes from interviews with Elaine and Brett in which they describe how they use their smartphones to immediately find answers to questions that they are concerned in the moment. These examples show that the awareness of having the ability to instantly "find out anything," as Elaine puts it, can change one's understandings of and relationships with the world, as well as the ways in which one sees oneself. Below, Elaine describes the situations when she prefers to use her smartphone to using a laptop or a tablet:

Elaine: I usually look on the phone, because it's so instant. I also use it if the children ask me questions I can't answer. I'll go straight to the Internet to find the answer …so, for instant answers to things, I just go on my phone.

Roxana: When would you do that?

Elaine: At any time. When I'm in the house, out of the house. When things just pop in my head. Lying in bed, going to sleep, "Oh, I'll just look at that up." It's brilliant; I wish we had that information when we were young. It's at your fingertips, really, to find out anything. (interview Elaine)

What Elaine describes in this extract can be regarded as a dialectic relationship that she has with her phone: any question that comes to her mind and that she addresses to the phone comes back to Elaine with an immediate answer in a never-ending exchange that can take place anywhere and anytime. The awareness that she can access this relationship anytime she needs to makes Elaine feel enthusiastic and confident in relation to her children—she can find the answer to any of their questions on the spot.

I cannot emphasize enough the enthusiasm expressed by the people who took part in my research in regards to situations when they were able to instantly reach the piece of information they needed. Ultimately, the aim of this chapter is to retain this enthusiasm and find and develop a suitable theoretical apparatus to express it to an academic audience.

A mother of two, Elaine got used to continuously swinging between working full-time, part-time, and taking longer periods of time off work in the eleven years since her first child was born. Her multiple interests and responsibilities, which she has learnt how to juggle simultaneously, are reflected in the types of questions that she uses her phone to find answers for. The ability to find out what she needs, when she needs it, makes everyday life easier. It helps Elaine stay in control and find solutions for various contingent situations, such as in the example below when she needs to decide on the right type of product for her daughter. There is a partial list below of pieces of information that Elaine searched for on her phone in the last two days before the interview was taken:

> I was looking at somebody else's website. I was translating 4 foot 10 inches into centimetres—I was in a shop buying some tights for Becky and she's between girl and adult … and I only know her height in centimetres, so I had to translate that. And then there was something medical—I was looking up about what exercises you can do. And I was looking at some holiday company. For some reason, I looked up the word "consensus," but I can't remember why. And I was looking at hairdressers in Middleborough because I had to find a new hairdresser. And then I was looking at jewellery boxes because I need to find a present for somebody. (interview Elaine)

This list, like any to do list, could be very telling of someone's everyday life and concerns. Using her smartphone, Elaine looks for relevant information that she momentarily needs to find in relation to her work, family, and, generally, to any of her various relationships and social roles. She is aware of her ability to use her phone to immediately find solutions to all the things that just "pop in" to her mind. I argue that this awareness, which Elaine enthusiastically shares with me, is empowering in itself. It makes Elaine feel able to respond to any contingent situations and, therefore, makes her feel in control of her life. At his turn, Brett, a father of two young children and a physical science academic, talks about a particular way of using his smartphone in order to check if what politicians say on TV is true:

> People say things on TV and I don't think they're right. So, I go and check them. Especially politicians ...it was on Question Time a couple of weeks ago. I can't remember exactly what it was, but somebody said something about how the deficit would change. So, I Googled it and looked at that. (interview Brett)

Brett describes his strategies for staying in control of the information that reaches him, of the information that he will believe. He is aware that the facts discussed by the media are not objectively true and just represent specific perspectives, which is especially visible in political discourse. Using his smartphone, he is able to immediately research other sources or perspectives on the same piece of information. By bringing all the perspectives together, he is able to reach his own conclusions, which are different from the conclusions offered by only one media source—the TV. Brett approaches various types of information, such as information on macroeconomic predictions, as being immediately available and not restricted to a small field of incontestable experts. Thus, by doing nothing more than using his phone skilfully while sitting on his sofa, Brett expresses himself as an agent who is able to contest political discourse and the power of government representatives. As for Elaine, the awareness of Brett being able to do so at any time and in relation to any potential piece of information is empowering in itself. It makes him feel in control of his own worldview, which cannot be easily influenced or manipulated by the media and by political discourse.

INTENTIONALITY AND WANTING AS PER ANSCOMBE

In *Intention*, the influential work that she published in 1957, the British analytic philosopher Elizabeth Anscombe patiently turns upside down all the existent assumptions about what intentions are and where they are to be found. One of her essential claims is that intention resides and appears with action and is not something premeditated or carefully constructed before the action starts. Anscombe suggests that we should look for intention not in the content of one's mind but in what one is doing. "Roughly speaking, a man intends to do what he does. But of course that is *very* roughly speaking. It is right to formulate it, however, as an antidote against the absurd thesis which is sometimes maintained: that a man's intended action is only described by describing his *objective*" (Anscombe 2000 [1957]: 45).

As Foucault was in his approach to the practical, which was discussed in the first chapter, Anscombe was influenced by Aristotle's work on ethics and praxis, and she developed a new conceptualization of the Aristotelian concept of "practical truth" (Anscombe 1981). The philosopher Roger Teichmann summarizes Anscombe's reconceptualization by suggesting that, for her, practical truth is when "what is aimed at by the agent is what he actually does: that the description of what he aims at is the same as the (or a) description of what he does" (Teichmann 2008: 81). In order to arrive at this formulation, Anscombe draws a parallel between "good action" and "true belief." She suggests that the "conceptual connexion between 'wanting' ... and 'good' can be compared to the conceptual connexion between 'judgement' and 'truth'" (Anscombe 2000 [1957]: 76) so that, as truth is the object of judgement, the good can be regarded as the object of wanting. In other words, as Teichmann puts it, "the question whether an action or end is in fact good is like the question whether a belief is in fact true" (2008: 74). This is an idea that, I believe, can find reflection in the discipline of social anthropology.[1] However, practical truth is something that concerns actions, not judgements. Following Aristotle, Anscombe sees "good action as embodying 'practical truth'" (Teichmann 2008: 79). The notion of good should be regarded here as relative; if the thing wanted has a "desirability characterization" given by the agent, then the thing wanted is "good in some way." "*Bonum est multiplex*: good is multiform, and all that is required for our concept of 'wanting' is that a man should see what he wants under the aspect of some good" (Anscombe 2000 [1957]: 75). To summarize Anscombe's reinterpretation of practical truth, I will use a quote from Teichmann's discussion of Anscombe's work: "Where what is wanted is wanted as good, and is in fact good, the action arising from the want embodies practical truth. Acting thus is, as Anscombe liked to say, 'doing the truth'" (Teichmann 2008: 80).

This philosophical understanding of immediate action relates to one of the benefits of spontaneity identified by the dramatherapist Peter Slade.

This is "a sincerity" about the way of doing a task (Slade 1968: 4). The sincerity in doing the task or, for Anscombe, acting as "doing the truth," is what differentiates spontaneous action from a "means-ends calculus" approach to action—or what the novelist Jane Austen called "acting by design"[2]—that characterizes "neoliberal agency" (Gershon 2011).

In an essay inspired by Anscombe's work, Charles Taylor (1979) approaches the question of practical truth using different wording. He writes about action as an expression of desire—what for Anscombe is "wanting"—and defines "happy action" as a qualitatively different type of action when "awareness of what I want is inseparable from awareness of what I am doing" (Taylor 1979: 86). Happy action is when the desire takes the form of unconstrained action; it is "a situation marked by an absence of conflict" (1979: 88). Happy action is a situation in which one is able to instantly do what one wants to do.

Wanting (or for Taylor, desire), thus, plays an important role in triggering spontaneous actions. But what is wanting? Anscombe finds it important to delineate wanting by outlining one basic characteristic, or "sign," of wanting that, as Teichmann suggests, would allow us to see "how wanting is different from wishing or hoping" (Teichmann 2008: 64). Developing her argument against the Cartesian account of first-person authority that reifies what Teichmann calls "'private', mental events" (2008: 17) and an interior causation for action, Anscombe insists that wants should not be seen as "inner causes." Instead, the way she approaches wanting is by reifying its "primitive" mark: "trying to get." Furthermore, she identifies two essential elements of wanting: a movement towards the thing wanted and a supposition that the thing is there:

> The primitive sign of wanting is *trying to get*: in saying this, we describe the movement of an animal in terms that reach beyond what the animal is now doing. When a dog smells a piece of meat that lies the other side of the door, his trying to get it will be his scratching violently round the edges of the door and snuffling along the bottom of it and so on. Thus there are two features present in wanting; movement towards a thing and knowledge (or at least opinion) that the thing is there (Anscombe 2000 [1957]: 68).

One could say that this description accounts for objects of desire that are material and reachable. An example from one of my interviews with Sam can help with the discussion of this case. In the previous chapter, I mentioned that Sam and her husband Peter have an allotment where they work every Friday morning from March to October.[3] Besides Fridays, they also go to the allotment spontaneously, when they realize that they have run out of vegetables to prepare for dinner. "We'll nip down if we just need to pick something for dinner. Like, say you want some carrots, or something, and you haven't got any," says Sam.

Compared to the routine Friday morning visits, these are unexpected trips that emerge from Sam's discovery that they have run out of vegetables. Going back to the two elements of wanting identified by Anscombe, the movement towards the thing wanted is exemplified here by the spontaneous journey to the allotment. While the supposition that the thing is there is expressed by their knowledge that they would find carrots hidden in the ground that they had previously planted and grown.

In situations when the objects of desire are less close and apparent, Anscombe regards the two features differently:

> But where the thing wanted is not even supposed to exist, as when it is a future state of affairs, we have to speak of an idea, rather than of knowledge or opinion. And our two features become: some kind of action or movement which (the agent at least supposes) is of use towards something, and the idea of that thing (Anscombe 2000 [1957]: 70).

At a first glance, this case could apply to spontaneous actions oriented towards finding a piece of information provisionally needed. I will look at this idea through an example coming from another participant, Matthew.

Matthew is in his fifties and lives together with his wife Carol and their two teenage children. He is passionate about history and enjoys doing Taekwondo together with his son and going to church every Sunday with his family. He is not very fond of new technologies—for example, he says that he prefers reading a paper book than using Carol's Kindle because he is "not good with buttons"—but he has an iPod that he uses at home with headphones to listen to audio books when it is his turn to clean the bathroom. Until very recently—a few months before the interview was taken—he did not have his own email address, so they used Carol's email to receive information about the church rota and other familial interests and activities. However, when I asked them during my interview if they ever used other digital devices while watching TV and in relation to the TV content, Matthew recounts an episode when he used his laptop to find information about the writer of a sitcom that he was watching:

The other day, *Red Dwarf* was on. And the first few series are written by two people, and the later are written by only one of them. And I thought, "What happened to the other one? Did he die or something?" So, I just Googled his name. He didn't die, he just didn't want to write it anymore. He was known as Rob Grant. And that's how I put it, and I found out. He said he wanted more than Red Dwarf on his tombstone, so he went on doing something else. So, it piqued my curiosity and I wanted to find out. (interview Matthew)

In this case, one could say that Matthew's drive to search for the information about the co-writer of *Red Dwarf* was not related to a skill or habit of mastering digital devices in a particularly technologically savvy way. Rather, it was a spontaneous wish related to what another participant defined as the cultural category of "is X dead?" questions. Thus, with his action, Matthew satisfies his immediate curiosity, which is linked to a cultural genre, through a new way of using his laptop in front of the TV. In this action, cultural continuity, technological change, and individual creativity are tightly intertwined.

Going back to Anscombe's theoretical apparatus, the two elements of wanting that can be identified in Matt's example are: (1) the action of writing down the name of the actor in the Google bar, which is some kind of action that the agent supposes is of use towards something and (2) finding out what happened to Rob Grant, whether he is dead and, if not, what the reason was for him stopping to co-write *Red Dwarf*, which is the idea of the thing needed. But if finding out what happened to Rob Grant is just the idea of the thing needed, then what is the thing needed? The main difference between Sam and Matthew's examples is that in one case the object of desire is material and edible (carrots for dinner) and in the other case the object is non-material or virtual (information about a name in the credentials of a TV series). The means of trying to get the objects are a combination of intense physical activities in one case—such as driving, digging and pulling out—and typing down a name in the other. But both approaches are based on the supposition that the thing wanted is there, be it inside the ground or on the Internet. It is with certainty that Matt Googles the name of the *Red Dwarf* co-writer. He is sure that he would find out at least if Rob Grant is dead, but he instead finds out more: the writer's reasons for stopping to work on the series. And he shares them with me as if we were talking about a common friend.

Another difference between Sam's and Matt's example might be the fact that digging out vegetables to have for dinner could be seen as a need, while looking up information about a sitcom writer would be a desire related to one's curiosity without being a need. However, I argue that the possibility of immediately accomplishing both of these types of needs or desires generates emotions associated with empowerment and self-confidence and brings out the opportunity for one to imagine oneself as an agent. Thus, for the purposes of the present chapter, the content of the desires is, per se, less important than the emotions generated by the acts of accomplishing them.

The next example compares the activity of physical search through encyclopaedias with the activity of searching on Google. Brett describes his in-laws' concept of the "pedant's corner," which institutionalizes a family tradition of what can be called friendly arguments. The pedant's corner is a bookshelf situated in the living room filled with encyclopaedias that are to be used spontaneously in case of unexpected friendly arguments over small but not unimportant issues:

> People get pedantic about something. And sometimes, in families, when you had a few glasses of wine, you can have an argument. Say, "Christian Slater was in this film." And then you say, "No, he wasn't in that film! He was in this film. In that film wasn't it Kevin Bacon?" You debate about that until you say, "OK, we'll check it!" So, a pedant's corner, in Jane's family, used to be a book shelf where they had things like encyclopaedias, and dictionaries, and the histories of people, and all that sort of things; so, very factual books. And, that's where they would go. But now, it's all in the phone basically. So, Wikipedia, really, it has become like the pedant's corner. Obviously, don't believe everything that's there … so, just in order to see who's right in a family discussion or argument. It's imperative that we have Internet signal. The worse thing is if you have these discussions and you are on holiday overseas and you don't get the Internet. Or, you stay somewhere in the country, in this country, and there isn't mobile signal. And then you have to write it down and check it out [later]. It's not something that rules our lives, but it's something that it's there in case we need it; the ability to prove yourself right or wrong. And, if you prove yourself right, then you tell everyone. And if you prove yourself wrong, then you don't say anything more about it. (interview Brett)

Here, Brett describes friendly arguments in families as an arena for spontaneous action that existed long before the development of new digital technologies. Friendly arguments are such a constant part of domestic life that his wife's family used to keep a whole bookshelf dedicated to these occasions. What is spontaneous is the subject of the debate—nobody can anticipate what the topic of this evening's friendly argument will be—and the moment when the opponents decide to check who is right, which happens after they present, more than once, all their arguments. During my fieldwork, I often took part in friendly arguments that usually happened in pubs when I was sitting with groups of local friends and acquaintances.[4] According to Brett's remark, friendly arguments happen today in similar ways to how they used to happen before digital technologies were widespread. What has changed is the research method for finding the answer. However, when new media fails to immediately provide the right answer due to technological constraints, people are left in limbo between the state of being right and the state of being wrong. To resolve this ontological dilemma in an everyday situation, people enact forms of ordinary agency that they have at hand and find appropriate. The question whether immediate media fulfil a need or a desire that is not a need is surpassed here; the ability to find a piece of information provisionally needed is equated with the ability to prove oneself right or wrong.

Summarizing Anscombe's approach to wanting, Teichmann concludes that "an object of wanting has to be taken as attainable by the person who wants it, and likewise has to be conceptualizable by that person" (Teichmann 2008: 65). This is the case in the three examples of Sam, Matt, and Brett: carrots for dinner, information about a sitcom writer, and the solution to a friendly family argument are taken as attainable (if the Internet is working). These examples also show that "trying to get—which for Anscombe is the primitive sign of wanting—is a feature present in both cases of a material and of a virtual object of desire, such as vegetables for dinner or a piece of information that would respond to one's momentary curiosity or would prove if one is right, respectively. I argue that, in these cases, new technologies do not alter the nature of wanting. They make a wider variety of wants achievable instantly. In Teichmann's (2008) terms, new technologies make new and diversified objects of wanting able to be considered attainable by the persons who want them.

I have discussed examples when spontaneity appears as an impulse to accomplish momentary wants. But there are also cases when specific wants appear first as elusive ideas that need to be accommodated until the stage when they will be considered full wants that ask for action. It was Chris, whom I introduced in the previous chapter, together with his family the Johnsons, who told me about this type of "fermented" spontaneity. In the next extract, Chris describes the way in which he reached the decision to buy a new game for his children and then used his smartphone to do so:

I bought a game of Cluedo. I was talking with the kids, or something. I must have seen it online, a reference to Cluedo, and then that made me think, "Actually, Cluedo's quite a good game for developing logical reasoning, and our kids are about the age where they can accept that sort of concept." So, that jumbles around in my head for a while. I'm the sort of person who, kind of, I like to allow things to bounce around in my head for a while and then it may look spontaneous from the outside, but it's been fermenting for a while. Then, all of a sudden, I'll go, "Aha, we need a game of Cluedo!" And, the problem with the modern life is that there's so little time to do anything, that I'm now in the situation where if I decide that I think I'm probably going to do something, at that moment I just say, "Well, I'm going to do it!" That's why I ordered Cluedo. I was, I don't know, I was in bed or something. I woke up and I thought, "Oh, I think we need Cluedo", and then it's done you see. It's just done. It's gone, without occupying any more of my mind, if you see what I mean. (interview Chris)

In Chris's case, spontaneity is put aside while the idea ferments in his mind. The moment when the desire reaches maturity is also the moment when it is fulfilled: once he realized that he wants to buy the Cluedo game, Chris immediately buys it using an online shopping application on his smartphone. He explains his action by stressing a lack of time as a characteristic of modern life and by suggesting that there are lots of other things that he is preoccupied with. If time is essential for an idea to mature, once the idea has become a decision there is no need to wait anymore in order to turn the decision into action. By acting spontaneously, Chris liberates the idea from his mind and makes room for other thoughts.

Going back to Anscombe's philosophical framework, one could connect her conceptualisation of wanting with her reconceptualization of the Aristotelian concept of "practical truth." As discussed, for Anscombe wanting involves two elements: (1) "trying to get," a movement towards the thing wanted which is spontaneous and almost concomitant with the occurrence of desire, and (2) the supposition that the thing wanted is there. In reconceptualising the concept of practical truth, Anscombe suggests that acting is "doing the truth"(Teichmann 2008: 80) because "the description of what he [an agent] does is made true by his doing it"(Anscombe 1981: 77). In order to connect these two ideas and express the significance of Anscombe's argument in only one sentence, one could suggest that in "trying to get" what they want, people are "doing the truth." This idea liberates any form of action from moral judgement, and, in doing so, I believe that it is congruent with the fundamentals of ethnographic description.

The ethnographic vignettes employed in this section reveal spontaneity as a social rather than an individual activity. A family visit to the allotment, having a friendly argument as a form of domestic sociality, buying a new game that attests the developing abilities of one's children, these are social preoccupations that are done together with or are oriented towards the others. The next section goes further by describing a few forms of sociality developed through the use of spontaneity and the ways in which the emotions generated from these processes contribute to broadening the ethical imaginations (Moore 2011) of the people involved, both participants and researcher.

SHARED SPONTANEITY AND ETHICAL IMAGINATION

I am hosting Annie, a couch surfer from the USA.[5] She came to Middleborough for a conference and had one extra day for exploring what she enthusiastically calls an "authentic and non-touristic" small English town. I am hosting her for one night in the Victorian terraced house that I share together with three friends. After showing her bedroom, we sit for a while in the living room, having a cup of tea and talking about ourselves. At one point, she mentions her favourite band, Bright Eyes, who are big in the USA, but that, I am ashamed to say, I have never heard of. Curious to find more about the band, I instantly download a couple of their albums on my phone through a music application. Then I ask what her favourite song is. I find it easily and press play. The song fills the living room. For me, new sounds are mixing with a familiar space, creating something different. For Annie, a familiar tune has just broken out, filling a strange place. We first met only two hours ago, but this is a moment when we kind of come together. We listen to the song and smile.[6]

Thinking back about this moment when I effortlessly enacted the spontaneity that my participants had talked about to me for over a year, I realize that it was Erin who taught me about instant gifting. I learned this form of sociality and exchange on a gloomy late August morning when I visited Joyce and Erin, one of the key participant families in my research, to give them a small souvenir that I brought from my holiday in France.

Compared to the four or five other times when I visited their place for specific research activities, there was no task we needed to complete this time. We were in the living room sitting around cups of tea when Joyce remembered the photos she took with me two months ago at Picnic in the Park when both Erin and I performed as part of the artistic programme, Erin walking on stilts and me Morris dancing. We were talking about these photos for a while. Joyce took them using the camera on her smartphone and then transferred them to my phone using Bluetooth. Erin has her own smartphone, which is a different and less expensive model than Joyce's and mine, but during the photo sharing we both consulted her as an expert. When Joyce and I finished with the photos, Erin used Bluetooth to connect her phone with mine. What could we share? We sat next to each other on the sofa. She was 11 and I was 29 and talked in a foreign accent, but we performed as part of the same parade a couple of months before and were now connected through Bluetooth. She asked me what my favourite animal is. "Well, I like dogs. My parents have a Labrador," I said. She turned to her phone, did something with it, and a new file arrived for download on mine. It was a cartoon of a very funny dog with its tongue out. We laughed together. Then I sent her a photo that I took of our dog Rich lying in the grass. "What's your favourite colour," she then asked. It is orange. Erin did something else on her phone, captivated and mumbling. Then a new file started to download on mine. What could it be? I looked at the blue downloading line growing and growing. I opened the file directly from the list of downloads. It was a beautiful orange tiger lily! I was so conspicuously impressed that Joyce, who returned from the kitchen with Erin's lunch, came to see what it was all about. I showed her the tiger lily, and she was also impressed. She did not know that Erin had such a picture on her phone.

During a subsequent visit, I asked Erin how she managed to send me pictures of my favourite animal and colour after I had just named them, like a fairy that could instantly make someone's three wishes come true. She did not have the photos on her phone, but she searched the words on Google Images, chose and saved the pictures on her phone, and then she transferred them to me. This series of actions were creatively employed together by Erin in that situation of exchange and of connectedness. They were not typical ways of using smartphones that one would be presented with in a user manual, for example, but they were adapted from previous actions of using Google on a computer and crafted together by Erin herself.

In this activity of crafting, Erin was aware of the capacity of the technologies she was using to provide immediacy. She had the knowledge that she would be able to instantly gift me with my favourite animal and colour if she only wanted to; in Teichmann's (2008) terms, Erin knew that her want was attainable. And by acting out this knowledge, Erin was extending her ethical imagination, the ways through which she imagined relationships with herself and with others (Moore 2011). Specifically, by proposing and engaging in a process of instant gift exchange, Erin created a spontaneous relationship between herself and a significantly different other. By surprising me and jazzing up a gloomy and uneventful morning, Erin was empowered to imagine that she would be able to recreate this form of sociality in other contexts and with various socially different others. I was empowered to think this as well even if I did not realize it until later after I enacted Erin's instant gifting with Annie. Thus, both Erin and I extended our ethical imaginations when we realized that we are able to spontaneously bring joy to significantly different others who, like Annie, were strangers one day before—just by using an ordinary digital device in a specific way.

Indeed, being able to achieve something instantly is valuable when it is a shared activity and ability. The people who took part in my research often described situations when they would sit together in the evenings in the living room around the TV, picking up facts that they would transform into common goals or friendly arguments, such as finding quiz show answers before the TV contestants. This is how Cynthia Smith and her daughter Kelly, who is in her early twenties, talked about the ways in which they collaboratively work to find out information about an actor that they vaguely recognize on TV:

Cynthia: Sometimes Kelly and I are watching TV, and there would be an actor. "Where do I know him from?" And, so, you pull up the information for that programme and you get the actor's name and you go to his page and find out what he was in. And you go, "Aha!" That sort of thing.

Kelly: Yeah. "Oh, what's his face, from what programme?" So, we have to look and find out who he is. (interview Cynthia and Kelly)

During one of my visits to Cynthia's place, I had the chance to better understand the joy of instantly finding answers to collective questions and of sharing one's findings with the others.

The living room was taken over by Kelly and her younger sister Emma who were making Wonder Woman costumes that they would wear at the jamboree they were attending with their guides group the following week. Kelly was already wearing a new blue skirt and a red top with the price tags still on while she was cutting white star shapes from an adhesive material sheet, which she would later iron onto the skirt. The screen of a laptop, placed on the living room table, showed a picture of the original Wonder Woman standing brave, hands on her hips, wearing her superhero costume. At the other side of the table, Emma was working in silence, absorbed by her own costume. I was sitting on the sofa with Cynthia, drinking squash and watching the girls. I asked her where the jamboree was held. It was in Essex. I knew this area well enough because a couple of friends live there and showed me many places around the county during several weekend visits. I asked where exactly in Essex. "Let's see," Cynthia said and lifted her own laptop from the floor with the enthusiasm of a detective who has been given a new mysterious case to solve. The laptop was already on so Cynthia just typed "Essex jamboree" into the Google bar, and she found the webpage of the campsite straight away. From this webpage, she clicked on the "How to find us" button and a map of the area instantly popped up on the screen. The camp was located west from Chelmsford and the names of the surrounding villages were not familiar to me because we mostly travelled towards north and east. Aware that her mum had the website of the jamboree on her laptop screen, Kelly started asking Cynthia questions about the camp's schedule, such as which themed party was planned for the first evening. Cynthia clicked, navigated, and found the answers for her daughter. Kelly listened without turning around, absorbed by her work at the costume. Cynthia put the laptop back down on the floor. Mission complete. Kelly then handed a white piece of paper in an almost-star shape to her mum, and Cynthia started trimming it carefully.

The laptop was used here as a way for Cynthia to participate in her daughters' preparations for the jamboree. She first attended to my momentary curiosity and later explored the website in response to Kelly's questions in anticipation of the jamboree that her daughters would take part in but that she would not. In this way, we all participated in the preparations for the camp: the girls by using a set of material objects and tools that they previously acquired specifically for the purpose of making their costumes, Cynthia and I by spontaneously using the laptop. We all shared in the anticipation of the event. This form of sociality was not planned, but it arose from the spontaneous use of the laptop in response to my question and from Kelly's spontaneous interest in response to Cynthia's navigation of the camp's website. Cynthia provided for her girls what they immediately needed and asked from her: information about the jamboree and help with trimming the star shapes. The spontaneous enactment of these actions of care and response generated agentic emotions. The awareness that Kelly's and my wants for information were attainable and that she could fulfil them in a moment using her laptop as a tool empowered Cynthia to extend a set of actions of care, which could be regarded as related to mothering, to others besides her children. She provided for me by immediately responding to my momentary curiosity related to the geographical location of the camp.

After this episode and after I visited them several more times (a couple of times for dinner) we planned a Saturday lunch visit when I would bring homemade lemon drizzle cake that I am particularly proud of. We set the meeting for 12:30 p.m. However, that morning Cynthia texted me asking to postpone the visit until 2 p.m. because Kelly asked her to join her at her hairdresser's that morning. I already had another appointment at 4 p.m. in a nearby town, so I texted her back suggesting that I just stop by to leave the cake, and that we arranged lunch for another date. Cynthia texted me back, "I can quickly feed you at 2 and we can plan a proper catch up time too." This suggestion was utterly comforting for an anthropologist when coming from a research participant. It made me think that Cynthia did not want me to travel to a different town without having had lunch and that she believed it her responsibility to feed me, as she already planned to do. I went, and we all had a nice quick lunch. This extension of actions of care that, arguably, started with interactions that involved the use of digital media and that developed with the growth of our relationship was an extension of both Cynthia's and my ethical imaginations.

SHARING BEYOND CO-PRESENCE

Another form of sociality that emerges from spontaneous employments of digital media devices concerns situations in which the participants in the exchange are not physically co-present. I will illustrate this form of sociality by discussing examples drawn from a research encounter with Lara, whom together with the Hewitt-Mitchell family, I introduced in the previous chapter.

I video recorded Lara for one of the project research activities, asking her about the ways in which she uses her smartphone. The photo camera feature of the phone seemed to be particularly important for her because Lara often used it to instantly "convert" material objects into digital ones, sometimes as part of an artistic process. For example, she takes photos of their own grown vegetables so she can return to the pictures when she needs to draw and create patterns for the pots that she makes during evening pottery classes. Every time she goes out, even just for walking the dogs, Lara remembers to take her phone with her so she can spontaneously take photos of and "keep" anything that grabs her attention, such as blossoming trees and green fields. The other day, she was looking through old photo albums and found a funny picture of her older daughter as a toddler. She took a photo of the material picture with her phone so that she would have it with her and be able to check it at any time, similar to the way in which people used to carry small pictures of their loved ones in their wallet. She also used her phone to post the photo on her profile on a social networking website, and her daughter, who is a friend of hers on this website, saw the photo and posted a comment. She showed me the picture while talking about it because it was so easy to instantly access it in the gallery section of her phone. Lara also used other social networking applications that focus on sharing photos to post pictures of her pots as well as some abstract and arty pictures that she shoots from unusual angles or in original compositions.

In her everyday life, while performing ordinary tasks and activities such as walking, cooking, or washing the dog, Lara's absent kin and friends sometimes appear in her mind. There are moments of joy, reminiscence, or ordinariness that she likes to immediately share with particular people, moments like her dog looking like a soggy sheep after having been washed:

> I was taking photos of Fizz this morning, because we had to wash her when we got home, and she was really soaking wet; she was like a little soggy sheep. So, I was thinking while taking the photos, "Oh, my mum would like to see this!" So then, I just emailed the photo to my mum, just straight away so she could see what we were doing this morning. (transcription from video recording with Lara)

In this example, Lara thinks of her mum through the photos. The multi-sensory experiences of her lived everyday life did not provide the space for Lara to imagine that her mum could be physically present in her house and, for example, help her to wash the dog. When taking the photos, the complex lived moment was transformed into an image that Lara was able to see in the same way that her mum would subsequently see it. The act of taking the photo, and Lara's ability to instantly send it, made Lara feel that she was sharing that moment with her mum even if they are not physically together. The mother was embodied in Lara's eye, which looked at the dog *through* the camera screen.

The domain of the visual can also bring out unexpected recollections for Lara. This happened when she played a guessing game based on drawing called Drawsome. Lara is used to playing this game on her phone with various people, such as old friends who now lived abroad, recent acquaintances living in proximity, and people that she does not know personally but who were proposed by the game. The game is played in pairs. One of the players is given a series of words from which they choose one that they will represent in a drawing and the other player tries to guess it. Lara described the situation when, while playing the game with an old school friend who lives abroad, she had to draw the word "tennis." This unexpected occurrence reminded her that she and her friend used to play tennis together, which made Lara write a note about this memory to accompany her drawing.

> She's somebody I went to school with; I've known her since I was seven, and now she's living in Australia. So, that's really bizarre. Sometimes you can talk on it; you can add a little note. So, one of the things she needed to guess was "tennis". And, when she was a child, they used to have a tennis court in their back garden. So, I said to her, "Do you remember we used to play tennis in your garden?" And, when it was her turn... you know, little things like this. (transcription from video clip with Lara)

It is the game that unexpectedly triggered the memory, similarly to the way in which a walk, for example, might serendipitously bring Lara and her friend in the proximity of a tennis court. By accompanying her drawing with a note concerning a specific memory, Lara redefined the drawing exchange in relation to the history of their relationship. She did not need to write a long letter with the specific scope of reconnecting with her friend. She could spontaneously bring out that memory through the game similarly to what would happen if, for example, she and her friend were taking a walk together.[7] At the same time, the opportunity to illustrate the word tennis to a friend whom one used to play tennis with as a child can bring back the memory of one's own childhood self. Lara was reminded that she had a history that might suggest a form of expertise for her in playing, talking about, and creating visual representations of the game of tennis.

These examples illustrate forms of sociality that could be described as virtual. They are articulated between people who do not share the same geographical location and are facilitated through internet-based communication—more specifically, through sharing images. What is different in these situations from other forms of communication between friends and kin living in different countries, which were sensitively documented in various studies on digital co-presence (e.g. Madianou and Miller 2011), is that, in the examples discussed, communication is not a planned and anticipated event as would be the case of a video chat that was arranged beforehand. Rather, it arises unexpectedly as part of everyday actions and routines. Because they are articulated through a combination of spontaneous usages of digital media, together with the spontaneous character of acts of recollection, the types of sociality that Lara formulates do not necessitate the responses of her mum or of her friend in order to emerge. For Lara, the forms of sociality are created in the moment when she sends the picture or the message, and they are still not one-sided. Rather than open-ended questions, her messages are instant gifts that Lara is able to imagine how the recipients will react to. The domain where these forms of sociality are formulated is, I argue, the domain of Lara's ethical imagination.

In describing a variety of situations in which she spontaneously uses her phone for taking photos of things and moments that she would like to remember and for sharing them with the others, Lara emphasized the importance of the visual in the way she expresses herself and communicates with others. I argue that, in this case, the photo camera on Lara's phone can be regarded as a tool for ethical work. Through the act of taking photos, Lara ordinarily articulates her perspective upon the world. She is aware that she can choose what to photograph and the perspective from which to regard the object from. She also knows that these choices are expressive of her subjectivity, of who she is, who she was, and who she would like to be: a potter, a gardener, a dog lover, a mum who has seen her now-grown daughter as a toddler, a childhood friend of someone who lives in Australia, a daughter who thinks about her mum who is in Spain. Past, present, and future are blended in the way in which Lara redefines herself, moment by moment, by taking new photos of what surrounds her. The "happy actions" (Taylor 1979) of taking and sharing photos at the moment when the wish arises are part of Lara's ethical practices.

Spontaneous Actions as Forms of "Doing"

Returning to the Aristotelian distinction between doing and making, a distinction that Faubion (2001) argues the philosopher Michel Foucault (1990, 2000) built his conceptualisation of ethical work upon, I argue that Lara's example and the overall ethnographic discussion developed in this chapter demonstrates that spontaneous actions are actions of doing rather than of making. The work of the philosopher Elizabeth Anscombe and her emphasis on the truth, or sincerity, inherent in acts of doing, provided important support for this idea.

In discussing Anscombe's argumentation in detail, my scope was to draw attention to a set of theoretical approaches that could provide real support for conceptualizing human action anew when looking at situations in which means and ends coincide—situations of doing. Besides the types of actions described in this chapter—those related to instant gifting, spontaneous collaboration towards finding factual information, and taking and sharing pictures—I believe that the theoretical apparatus proposed here could be applied and advanced in research looking at various aspects of life and forms of activity that might represent doing rather than making, such as sports activities, performing arts, hacking (Coleman 2013), flash mobs and various other forms of protest, human-animal relationships, countryside walking, or play.

Notes

1. Joel Robbins (2013) proposes to move towards an anthropology of the good to recuperate the critical force that the notion of difference, and the process of learning from alterity, have in social anthropological analysis.
2. "'Your plan is a good one,' replied Elizabeth, 'where nothing is in question but the desire of being well married; and if I were determined to get a rich husband, or any husband, I dare say I should adopt it. But these are not Jane's feelings; she is not acting by design'" (Austen 1993: 20).
3. Two interesting approaches to gardening in England can be found in Degnen (2009) and Tilley (2006).
4. An example of memorable friendly argument that my local acquaintances and friends engaged in while in a pub was over the origins and correct pronunciation of the word "mum."
5. Couchsurfing.com is an international online network for hospitality.
6. Bialski (2012) defines the interactions between couch surfing hosts and surfers as "intimate mobility".

7. Paula Uimonen (2012) uses the concept "transtemporal" to define the way in which her participants used Facebook to maintain relationships with people from different stages of their lives.

References

Anscombe, Gertrude Elizabeth. 1981. *From Parmenides to Wittgenstein: Collected Philosophical Papers of G.E.M. Anscombe. I.* Oxford: Blackwell.
Anscombe, Gertrude Elizabeth. 2000. *Intention.* Cambridge, MA: Harvard University Press.
Austen, Jane. 1993. *Pride and Prejudice.* London: Wordsworth Classics.
Bialski, Paula. 2012. *Becoming Intimately Mobile.* Frankfurt am Main: Peter Lang.
Coleman, Gabriella. 2013. *Coding Freedom: The Ethics and Aesthetics of Hacking.* Princeton: Princeton University Press.
Degnen, Cathrine. 2009. "On Vegetable Love : Gardening, Plants, and People in the North of England." *Journal of the Royal Anthropological Institute* 15: 151–167.
Faubion, James D. 2001. "Toward an Anthropology of Ethics : Foucault and the Pedagogies of Autopoiesis." *Representations* 74 (1): 83–104.
Foucault, Michel. 1990. *The History of Sexuality,* vol. 2. London: Penguin Books.
Foucault, Michel. 2000. *Ethics: Subjectiviy and Truth (Essential Works of Foucault 1954–1984).* London: Penguin Books.
Gershon, Ilana. 2011. "Neoliberal Agency." *Current Anthropology* 52 (4): 537–555.
Green, Sarah. 2002. "Culture in a Network: Dykes, Webs and Women in London and Manchester." In *British Subjects: An Anthropology of Britain,* edited by Nigel Rapport, 181–202. Oxford: Berg.
Hochschild, Arlie. 1997. *The Time Bind: When Work Becomes Home and Home Becomes Work.* New York: Metropolitan Books.
Luhrmann, T. M. 2006. "Subjectivity." *Anthropological Theory* 6 (3): 345–361.
Madianou, Mirca, and Daniel Miller. 2011. *Migration and New Media: Transnational Families and Polymedia.* London: Routledge.
Mauss, Marcel. 1979 [1935]. "Body Techniques." In *Sociology and Psychology: Essays by Marcel Mauss,* translated by B. Brewster, vol. 2, 95–123. London: Routledge & Kegan Paul.
Moore, Henrietta L. 2011. *Still Life: Hopes, Desires and Satisfactions.* Cambridge: Polity Press.
Robbins, Joel. 2013. "Beyond the Suffering Subject : Toward an Anthropology of the Good." *Journal of the Royal Anthropological Institute* 19: 447–462.
Shove, Elizabeth. 2009. "Everyday Practice and the Production and Consumption of Time." In *Time, Consumption and Everyday Life,* edited by Elizabeth Shove, Frank Trentmann, and Richard Wilk, 17–34. Oxford: Berg.

Slade, Peter. 1968. *Experience of Spontaneity*. London: Longmans.

Taylor, Charles. 1979. "Action as Expression." In *Intention and Intentionality: Essays in Honour of G.E.M. Anscombe*, edited by Cora Diamond and Jenny Teichmann, 73–89. Brighton, UK: Harvester Press.

Teichmann, Roger. 2008. *The Philosophy of Elizabeth Anscombe*. Oxford: Oxford University Press.

Tilley, Christopher. 2006. "The Sensory Dimensions of Gardening." *Senses and Society* 1 (3): 311–330.

Uimonen, Paula. 2012. *Digital Drama: Teaching and Learning Art and Media in Tanzania*. London: Routledge.

CHAPTER 6

Anticipation and the Mother-Multiple

After lifting a large plastic basket filled with unpaired socks from the floor to the top of the dining table, Cynthia looked into the camera.

"And, again, is you who sorts the socks?" I asked, holding the video camera.

"Generally, yes. Again, sadly, my brain knows exactly whose socks are whose, so I can do that. Whereas, if somebody else pairs them, I then have to take them apart again to see whose they are. So, it tends to be my job," Cynthia said amused.

It was Sunday morning, and I was visiting Cynthia and Jeff's place to film and ask them about how they do their laundry. After showing me upstairs her method for sorting the existing laundry into specific piles for washing, Cynthia talked back in the living-room about an unfinished job that she did not manage to do the previous week: sorting out the family's socks. These are the socks that were washed the Sunday before and were waiting to be identified, paired, and returned to their owners. She usually did this in the evening on the sofa while watching TV, but last week she was too poorly to engage with the contents of the socks basket—a considerable content of probably over 40 items belonging to five different people.

"We've got a lot of black socks, and they are all pretty much the same size. I know that one's mine and that one's Kelly's; that one's Jeff's. And these ones are Lee's," Cynthia said as she picked the items one by one and

© The Editor(s) (if applicable) and The Author(s) 2016
R. Moroşanu, *An Ethnography of Household Energy Demand in the UK*, DOI 10.1057/978-1-137-59341-2_6

aligned them on the table. They all looked similar to me. I hide behind the video camera, which filmed these personal items that belong to people who were taking part in my research. Meanwhile, Cynthia stood tall, digging into the basket confidently to pick, name and manipulate the socks. She welcomed every finding with recognition, giving them the name of one of her loved ones.

While she regarded her knowledge of the items as being permanently stored on her brain, this knowledge needed to be enacted, as in the practice of pairing socks, to keep the household on track. While managing the content of the socks basket, Cynthia gave my video camera and me a glimpse of this enactment that was part of practices of care giving. I propose to regard such moments and practices, when one purposely acts as the caregiver of one's "domestic others"—family members, pets, or the home itself as an entity—as moments of being the Mother-Multiple.

The Mother-Multiple describes a mode of being that any individual can access when engaged in kinship-situated caring. As such, it is an ontological position that is enacted in specific practices that are deliberately oriented towards one's domestic others and involves the anticipation of their needs, habits, preferences, and dislikes, such as the act of preparing a cup of tea for one's partner exactly the way he or she usually takes it. My use of ontology here follows an approach to ontological multiplicity as practical, as emerging from distinct enactments similar to the way in which Mol (2002) looked at ontological multiplicity in her work on medical practices of diagnosis in a Dutch hospital.

While the purpose of the concept Mother-Multiple is to open up kinship-situated caring to anyone, regardless of their gender, age, and child-bearing status by including the term "mother" in its naming, I want to acknowledge the close association between caring and mothering in Euro-American contexts (Bowden 1997).[1] Moreover, as it is a concept developed from ethnographic work with and for the purposes of looking analytically at English middle-class families, I retained the figure of the mother as traditionally associated with a series of duties of domestic improvement—of one's children and of one's domestic space—as this idea emerged with the appearance and the development of the middle classes in the UK in the nineteenth century (Strathern 1992: 104). Therefore, the concept of Mother-Multiple should be regarded as emerging from a specific ethnographic context. One of the ways in which the expectation for a mother's preoccupation with the wellbeing of her domestic others continues to be expressed today in this context is through the expression "being mother" that refers to the act of serving food or serving tea from a teapot to the others. Just as anybody at the table could provisionally "be" mother, I argue that any family member can and does enact the Mother-Multiple in specific situations and at different moments of the day.

The Mother-Multiple involves an irreversible connection, such as kinship relations in an English middle-class context are considered to be (Strathern 1992), that finds a suggestive expression in the following quote: "I thought oh God: I've got her till I die–it was this attitude–even when she's away married and all the rest of it I'll still worry about her" (Oakley 1979: 143). The vivid realization of this irreversible kinship connection belongs to one of the participants in Oakley's (1979) study of first-time motherhood. What the participant expresses here is the realization that from now on and until the end of her life, she would not be able to *be* anymore without carrying the thought of her daughter's existence in her mind. The participant suddenly finds herself in a different ontology than the one she occupied before her daughter's birth and the one her childless friends might still occupy.

The way in which the Mother-Multiple is enacted in English middle-class kinship relations refers to incorporating one's domestic others. For example, when being Mother-Multiple, the thoughts that one has of the existence, of the character, habits, needs, preferences, and dislikes of the others become part of one's everyday actions in a way that makes it hard to disentangle oneself from the individual others, just as Oakley's participant sees herself as being preoccupied with her daughter at the same time as with herself. This entanglement of individuals is manifest in Cynthia's socks basket. Being in charge of the basket is being the Mother-Multiple: a way of operating not from the standpoint of an individual, but through accounting for a set of relationships.

DIVIDUAL/INDIVIDUAL AND KNOWING OTHER PEOPLE VERY WELL

The Mother-Multiple can be described as the state of *being by and through a set of relations/connections* while the ontology of the individual, in an English middle-class context, can be regarded as *being by its own/in itself*. I build this distinction by drawing upon Strathern's (1992) argument that in middle-class English kinship, the individuality of persons is prior to the relations that bring them together. This observation emerges in relation to Strathern's previous experience of conducting ethnographic research in the islands of Melanesia and her development of an analytical framework for regarding persons as being conceived both individually and dividually. The Melanesian notions of personhood differ from Western ones in their focus on a relationally composition. Melanesian persons are made up of multiple beings refracted through multiple relationships developed over time and starting from their procreation (Strathern 1988). In relation to gender, it is shown that, as deriving from different-sex parents, the child combines within itself both female and male elements. Therefore, in the Melanesian context gender is not employed to express the identity of whole persons as in Western contexts, but gender differentiates types of sociality like same-sex bonds from cross-sex bonds (Strathern 1988: 324). Dividuals are, therefore, persons that are constructed "as the plural and composite site of the relationships that produced them" (Strathern 1988: 13) and change over time by developing more relationships. Meanwhile, individuals in middle-class English kinship are conceived as existent and finite persons that are prior to relationships.

I argue that the Mother-Multiple is neither a dividual nor an individual because it is not a person. Rather, it is a state of being that is accessible to individuals and essentially provides them with a change of perspective. This is the perspective of the multiple relationships with their domestic others rather than the perspective of one individual or another. The Mother-Multiple is a knot where different kinship relationships intersect; it manifests the agency of this intersection of relationships rather than the agency of one or more individuals. A good visual representation of the Mother-Multiple as a knot of relationships is Cynthia's basket of socks. As a representation of the Mother-Multiple, the basket contains all the family members and the relationships between them. Whoever takes the basket in order to engage with its content is enacting the Mother-Multiple ontological position. They momentarily stop operating through a set of individual lenses, such as the wish to not know which socks are whose in order to retain more important information in one's mind, and operate through the perspective of a knot of relationships.

The way in which I draw here upon the distinction between the individual and the dividual proposed by Strathern (1988) is by suggesting that while the main principle in constructing personhood in middle-class English kinship is individuality, when individuals engage in actions of caregiving it is still possible for them to access and to momentarily inhabit a relational and multiple perspective that is similar to what the dividual is in a Melanesian context. The notion of dividuality is employed here as a conceptual tool that can be used to shed light upon a specific process in Euro-American contexts, similar to the way in which Simpson (1998) employs this concept in his work on divorce in the UK. Simpson (1998) suggests that situations of divorce can make visible an approach to children as dividuals. When partners live separately, with children shifting between the two places, the influence of one parent and their new family or the extended family upon the children's upbringing becomes an issue that is detectable and can be upsetting even if it is considered inevitable. In these situations, therefore, children are regarded as dividuals, composite persons made of parts that correspond to the different relationships that they are engaged in, rather than as individuals.

One of the reasons why people enjoy being the Mother-Multiple from time to time is, I argue, an interest with knowing other people very well. The Mother-Multiple ontological position provides an epistemology consisting of special knowledge of the character, habits, needs, preferences, and dislikes of one's domestic others, as well as of the relationships between different domestic others and between domestic others and oneself. Expressing this epistemology is a form of agency. By showing that they know the others very well, people also express something about themselves. They prove their capacity to tell how the others really are, a capacity Rapport (1994) argues can involve multiple strategies and long-term training.

Several social anthropologists doing fieldwork in various parts of Britain have identified the interest with having profound knowledge of other people as being an important preoccupation of their research participants in numerous contexts. Cohen (1987: 70) compares what knowing someone means in an urban setting (to be acquainted with someone) with what it means in Whalsay, a small town in the Shetland Islands. In Whalsay, where everyone is already acquainted, to know someone means to know of their connections through, for example, families, boats, and houses. Edwards (2000) cites Cohen's comparison in her work on kinship and new reproductive technologies in a Lancashire town when she analyses the statement of a participant who outlines that she could never "fully know" her two step-grandchildren because she does not know their father's family. Edwards outlines that "[t]o know her grandchildren fully Mrs Watson suggested, required a knowledge of those to whom they were connected. This is what it means to be *close*'" (Edwards 2000: 245). In this case, even if she would have had the capacity to do so, the possibility for knowing her grandchildren fully is denied to Mrs Watson by the circumstances.

In the case presented by Rapport (1994), the capacity to tell the real character of others is necessary in Wanet, a small Cumbrian village, in relation to the changes brought by the appearance of outsiders who express worldviews foreign to Wanet. The worldviews of these outsiders, or "off-comers, come from an intellectual system of knowledge based upon words and imagination rather than from the practical knowledge of working the land and making a living that is specific to Wanet. A series of clues are brought together to know people for what they truly are: their horoscope signs, the biology of their parentage, details of their family history, knowledge of their behavioural patterns over the years, and their body language (Rapport 1994: 232–236). This complicated series of interpretative acts need to be performed because it is believed that there is a difference between how people appear and how they truly are—between words and "a deeper reality" (Rapport 1994: 225). This difference, in the novelist's E.M. Forster words employed by Rapport, is between the "two kinds of human personality which an individual possesses" (Rapport 1994: 220), a public personality and a private, lower, and deeper personality that transcends the everyday.

The ideal of being able to profoundly know someone other than oneself is described by Reed (2011) in his ethnography of fiction reading with the members of the Henry Williamson Society. For Reed's participants, reading Williamson's captivating novels brings no less than the experience of being embodied by the writer's consciousness: "For the first time, Williamson readers claim to experience a person from the inside out, to live, as opposed to guess or interpret, alien character and intentionality. What they believe they achieve is an extraordinary, previously undreamed sense of intimacy with another subjective mind; readers know Henry in a way they cannot possibly know anyone else" (Reed 2011: 10). Achieving this form of intimacy with another person's mind is valued by Williamson's readers because this is not an ordinary experience and cannot be replicated even with one's closest others. In order to maintain this intimacy, readers collect Williamson's novels because every book means a new encounter with what for them is a well-known consciousness. As Rapport, drawing upon the work of E.M. Forster, suggests, "it is from Literature that one receives hope of knowing others as one knows oneself" (Rapport 1994: 244).

While this hope might indeed come from reading fiction, the preoccupation with knowing others as one knows oneself is frequently expressed in everyday domestic situations between kin. Besides activities related to doing the laundry that were mentioned earlier, another domain where this preoccupation is visible is cooking.

Joyce is vegetarian and enjoys cooking from scratch for herself. She also hopes to convince her daughter to adopt some of her favourite dishes and give up the semi-prepared food that she prefers, such as fish fingers and chicken teddies. Joyce knows that this transition would only happen over time and is patient for it to emerge, paying attention to any new changes in her daughter's taste and supporting them. While filming her cooking carrot and lentil soup, Joyce told me with excitement that the last time she cooked this dish her daughter liked and ate it. She was now trying to replicate the exact way she prepared the recipe last time, which involves cooking the ingredients separately and not adding spices, so that her daughter will like it again. However, because she knows Erin very well, Joyce knows that liking and not liking specific types of food also depends on the frequency of having them. Once ready, she divided the large quantity of soup into portions and froze them so she could subsequently serve the dish to her daughter not more often than twice a week during the following weeks. When she tried the soup, Joyce could tell how it tasted for her, but she also tried to anticipate and imagine how it would taste for Erin. By stepping into the Mother-Multiple position, Joyce tried to access her daughter's sensorium—she attempted to incorporate the ways in which Erin sees and senses the world. The actions of cooking express the relationship between them, between the mother's individuality and the daughter's individuality that come together in this practice in a distinct way. The enactment of the cooking activities is done by Joyce, but the sensorium that is (imaginatively) engaged is Erin's. This mixture emerges from Joyce's interest with knowing her daughter very well so she can help her improve her diet, an interest that she pursues by enacting the Mother-Multiple ontological position.

"Knowing of" in Ethnographic Research and in Being Mother-Multiple

The affective qualities associated with domestic settings and to family-style lifestyles, which refer to a specific intensity of both positive and negative emotions (Gabb 2010), need to be accounted for when looking at everyday domestic practices such as cooking and doing the laundry. For example, as I have found out myself, the action of washing dishes feels very different when it is performed in a commercial or professional setting than it does in a family-style setting. When attending a large music festival during the year of my fieldwork, I volunteered to work in a cafe in exchange for a free ticket. One of the tasks I carried out on this occasion was to wash dishes for over three consecutive hours. I did the task without paying much attention to my actions, handling the piles of anonymous plates absently while chatting with another volunteer. By contrast, in a family-style lifestyle setting, washing dishes is, for me, an encounter with the plates that have been used by my domestic others. I am able to tell which plate was used by whom because I know their food preferences and just had the meal together. These are not anonymous pieces of crockery. They are objects that people I care for have used to feed themselves and stay alive.

By suggesting that actions of caregiving in domestic settings come with nothing less than the opportunity to step into a distinct ontological position, I attempt to preserve and express the affective complexity of domestic universes that I encountered during my fieldwork. Surely, a different eye could look at my ethnographic material from the perspective of, for example, existent sociological concepts in the field of family studies. By following an anthropological endeavour of ontological orientation (Henare et al. 2007), however, my analytical focus is not upon explaining empirical data by using existent theoretical frameworks. It is upon the possibility of creating new theory from ethnographic insights, which is an endeavor that necessitates a work of assertion, insistence, and imagination.

In writing this chapter, more so than in writing the other chapters of this monograph, I counted upon a distinct form of knowledge that I acquired as an ethnographer. Luhrmann (1989) calls this type of knowledge "knowing of," and she defines it as the intuitive knowledge of just "grasping" meanings, the key feature of which is an "imaginative absorption with a different person or world view" (Luhrmann 1989: 203). This type of knowledge, which Luhrmann describes in relation to neo-pagan witchcraft practices in the UK, complements the traditional distinction drawn by Ryle (1949) between "knowing that" (propositional knowledge) and "knowing how" (embodied knowledge). As opposed to propositional knowledge, which is achieved in all social science disciplines through the method of interviewing, knowing of is one of the main types of knowledge that the ethnographer gains by doing long-term fieldwork.

From all the ethnographic relationships that have emerged during my fieldwork, the relationships with the female adult participants played a special role in how I came to look at my findings. This was not only because of similarities in terms of gender and age that made it easier for us to get closer, but also because of one main difference and my interest in describing it. My participants were mothers, which I was not. Some of them were homemakers and some of them worked in part-time or full-time employment, but regardless of this difference, they all engaged in child-related "duties"—a set of practices and activities that I was not familiar with as an adult even if I could remember that I was frequently a child-minder for my younger brother when I was a child. As an adult in my late twenties to early thirties, equipped with critical thinking and having engaged with feminist literature and artistic products, there was also a personal scope when I tried to understand the lives of my female adult participants. I hoped that the knowledge gained from these relationships would allow me to tell if "it would be worth it" for myself, to know if I wanted to become a mum someday and if I was equipped with the necessary traits that would make this a positive experience for all those potentially involved. This personal concern might have made the exercise of reaching an imaginative absorption with the lives and worldviews of my participants who were mothers a more accomplished endeavour than the acts of imagining the lives of the male adults and the children participants.

Besides being a form of knowledge that is specific to ethnographic ways of learning, knowing of (involving imagination, intuition, and an interest with knowing someone very well) is a type of knowledge that people employ in everyday life in relation to their domestic others. Knowing of is one of the forms of knowledge that can be accessed when one steps into the Mother-Multiple ontological position, together with propositional and tacit knowledge. In the example of Joyce cooking soup for her daughter discussed earlier, Joyce employed all these forms of knowledge. Propositional knowledge was expressed when Joyce explained her strategies of dividing the soup into portions and freezing them so that she could alternate Erin's food menu, which was originally a technique explained in parenting literature on ways of diversifying the alimentation of one's baby. Tacit knowledge emerged in relation to the activities of cooking that Joyce performed "skilfully" (Grasseni 2007) because her body knew their sequence as it would know the steps of a dance. Finally, Joyce employed a knowing of form of knowledge when she imagined and accessed her daughter's sensorium in the act of tasting the soup as Erin would.

SHORT-TERM ANTICIPATION AND ENERGY CONSUMPTION

Anticipation of the needs and reactions of one's domestic others are part of practices of caregiving. When employed in this way, as directed towards one's domestic others, acts of anticipation can be regarded as acts of care in themselves. The concepts of Mother-Multiple and anticipation, therefore, are described in this chapter as shedding light upon one another. This outcome emerged from my fieldwork, during which activities of anticipation were described more often in relation to being Mother-Multiple than in other situations.

I regard anticipation here as one of the temporal modalities developed and enacted in domestic settings that engenders particular forms of ordinary agency. The way in which anticipation as a time-mode blends the present and the future, in creating a counter discourse to linear time, is by covering the present in the imagination of the future. This makes the future present. As part of everyday life, anticipation takes two forms, short-term anticipation and "looking forward to." Short-term anticipation emerges from thorough knowledge, based on previous experience, of how someone or something is likely to act.

During a video tour of her house, Cynthia told me "the hall light only goes on if somebody's out and we're expecting them back." As the light switch in the hall is on the opposite wall and not by the front door, a family member who goes out and is due to return later—usually Kelly, the oldest daughter in her early twenties—would need to step in a dark environment and find her way to the opposite wall to switch the light on. This might be an even more difficult action when someone is returning after a night out. Cynthia said they would leave the light on until Kelly returns "for accident prevention: it's a lot safer to come to a lit space than to a dark one." But it is also because "it's a welcoming thing I suppose. 'We're all expecting you back, don't disappear!'" Cynthia leaves the hall light on in anticipation of her daughter's return. The light is not only a sign for Kelly that she is expected back, but also a means for Cynthia to tell herself that Kelly is due to return. The light makes the mother's anticipation visible. It also shows Cynthia's embodied knowledge of her home and the material dispositions of walls and light switches, which might be based on previous experiences of being the first one to arrive home and finding the hall dark.

This type of anticipation—based on tacit, embodied, practical knowledge—appears very often when people are engaged in complex activities that involve multiple appliances and material objects, such as in the process of cooking. In these cases, ovens and kettles, for example, are switched on well before they are needed because people know they will use them at some point in the process of cooking and prefer to have them ready rather than wait for them to warm up at the last minute. "Let me put the kettle on" was how some of my key participants welcomed me several times towards the end of my fieldwork. And with this line we would anticipate together a nice catch-up over a cup of tea before I would go on with the interview or with explaining the task that I was proposing that time.

The downside of foresight in relation to the process of cooking is the additional energy that is consumed pre-emptively in expectation. If the kettle has boiled before the vegetables are peeled and chopped, it would need to be switched on again when the vegetables are ready to go in the pan. In these situations, being prepared to carry out the process of cooking with no interruption takes precedence over being energy efficient. This attitude of foresight in the kitchen is, as I discovered, part of the ways in which food recipes are explained in the UK. When making a cake from a British recipe for the first time, I was surprised to find out that the very first instruction was to switch the oven on so that it would reach the desired temperature by the time the cake mix is ready. By following these instructions step by step, I discovered more than once that either my oven was too fast to reach the desired temperature or I was too slow in preparing the cake composition. Either way, unnecessary energy was consumed during my training to become an agile baker just from foresight.

Short-term anticipation is also expressed in activities of setting up an environment, such as leaving a light on after preparing the dining table in anticipation for the family members to gather and dinner be served or switching the radio on to prepare the kitchen—and oneself—for starting to cook. During an interview about digital media and domestic time, Sam mentioned that she and her family liked listening to music while they have dinner, and she described how her husband and daughter picked the CDs to be played at dinnertime:

> Or when we're all around the table he would perhaps say, "Right, what kind of music would you like?" And then he and July usually tend, between them, to choose, right before we sit down. So, yeah, it's kind of home when the meal is served up and he starts picking the music. (interview Sam)

In this example, picking the music anticipated the experience of having a nice family meal. For Sam, their house started to feel like home before they sat down when her husband started to pick the music. Thus, it was not only the family dinner as event, but also the short-term anticipation of the dinner that covered and extended the event, that triggered the feeling of home for Sam.

In the examples discussed, by expressing the expectation of a series of actions that will follow a specific moment, such as using kettle-boiled water in the process of cooking or knowing that everybody will take a seat around the table once the music starts playing, the participants showed their knowledge of their domestic environments, which were populated by domestic others and defined by specific ways of performing everyday practices. When revealing themselves as experts in the domain constituted by their own domestic environments, and when enacting the exact activities that needed to occur at specific points in the progress of practices to keep the flow going, people manifested agency.

What happens in one's mind when engaged in anticipation can be regarded as specific techniques of imagination. In her work on neo-paganism and witchcraft in London in the 1980s, Luhrmann (1989, 2002) identifies two main techniques that magical practitioners used to access the "otherworld" and to train their capacity for "dissociation" (2002): meditation and visualization. Meditation is a concentration technique that is widespread in various spiritual traditions (see Cook [2010] for an ethnographic account of Buddhist meditation) and has been recently researched and reintroduced to Western practitioners under the name of mindfulness meditation by self-help programmes such as mindfulness-based stress reduction (Kabat-Zinn 1994) and mindfulness-based cognitive therapy (Williams and Penman 2011). Visualization, in Luhrmann's work, is a technique of imagination of a still mental-image or of "pathworkings": guided exercises of visualization of a fantastic journey taken by the practitioner, often conducted in groups, with one person leading by talking the others through the stages of the journey. As Luhrmann (2002) remarks, the capacity for visualization can be trained and is used at the moment particularly in performance sports training with techniques called "guided imagery" and "visual rehearsal" (Ungerleider 1996). Guided imagery is oriented towards achieving an internal visual representation of the sensory experience of performing a sport so that athletes become even more intimately familiar with their sport by creating a personal "image bank" that could be accessed anytime. Visual rehearsal, however, is focused on a specific task, such as the preparation for competition, with the athlete imagining in detail and in slow motion his or her body movements for a successful performance. Visual rehearsal can also be used during the game, just before an important moment like a decisive kick.

If imagination techniques are used self-consciously as tools to improve performance in sports training, their everyday employments are often dismissed or downplayed because, as Ehn and Löfgren (2010) show in their work on "non-events" (activities of waiting, routines, and daydreaming), they might be considered "either too ordinary or too insignificant" (Ehn and Löfgren 2010: 4). At the same time, they are part of a dimension of inner life or interiority, the accessing of which has not been a traditional preoccupation of social anthropology (Irving 2011). Ehn and Löfgren argue that inner life, or what they call the "secret world," is shaped by social life as much as any other human experiences or actions: "Events we believe we have invented, rules we believe we have created ... all turn out to be shared with others—but in secret" (Ehn and Löfgren 2010: 3). Therefore, it is possible that techniques of imagination similar to those used in sports training to improve performance are also employed in everyday life in relation to domestic activities. I propose to regard short-term anticipation as a form of visual rehearsal. By doing so, the act of putting the kettle on can be regarded as involving the imagination of the series of actions that would follow in the process of cooking. Similarly, looking forward to can be considered a form of "non-guided" imagery.

Looking Forward to and the Agency of Imagining Possibilities

Looking forward to involves the anticipation of an event that is not immediate. It can refer to cyclical occurrences, such as recreational or hobby-related activities (gardening sessions on Friday afternoon and Morris dancing practice on Sunday evenings were my own looked forward to activities during my PhD), or it can refer to larger events, such as Christmas, one's birthday celebration, or a holiday.

When meeting Sam once at the end of November so I could buy a few pairs of earrings made by her daughter Julie that I wanted to gift for Christmas to my friends in Romania, we chatted at length about Christmas traditions. She said that they always decorate the Christmas tree on the 1st of December because they like to keep the tree for an entire month because they actually prefer "building up to Christmas" to Christmas itself. The period when you prepare for it to happen, wait for it, and think about it is more valuable because "when Christmas comes and it's there, next thing it's gone and it's nothing else left to expect" (extract from my field notes). From the 1st of December until Christmas is twenty-four days of anticipation. All of December, as a time that leads up to Christmas, is a form of "building up"—a liminal time of imagination and of "ontological security" (Giddens 1991) that one reaches by placing oneself in the cyclical form of temporality associated with a traditional, stable, annual event—that is regarded as being more enjoyable than the event itself.

The technique of imagination that one employs when looking forward to is a form of "non-guided" imagery. It is self-generated thought about a future event. Through imagination, the event is internally accommodated and domesticated; one has the opportunity to get used to the idea of Christmas before the celebration would take place.

Having something to look forward to is a general concept that is understood in other domains of life as well. While taking part with my Morris side in an Area Community Forum event regarding the distribution of small grants to community projects, I encountered a different way of using the notion of looking forward to. The event consisted of a set of projects competing for the bid being presented by their proponents to council representatives and to local residents, followed by the audience marking each project. The president of a community association introduced their project for "fit bodies and minds," which involved organizing sport activities and quiz evenings for the residents in their council estate. She described what everyday life could be like for her neighbours, who were living in poverty and/or with disabilities. She said that their proposed activities would make everyday life more bearable because they would provide the residents with "something to look forward to." "We all need something to look forward to," she said at the end of her presentation, while the audience nodded in silence in understanding and agreement.

In writing about time, the sociologist Barbara Adam (1995) outlines that we are always engaged with a multitude of times and temporalities, such as clock and calendar time, the time of the body/biological rhythms, the cyclical time of nature and seasons, a past time that we access through memories, and an imagined (hoped for, or feared) future. All these times and many others coexist, and people often employ them simultaneously to make sense of their experiences. Waking up in the morning in pain and in a cold house that one cannot afford to heat, and knowing that nothing is due to happen on that day is different from waking up in the morning with the anticipation of an afternoon meeting and a quiz at the community centre.[2] Having something to look forward to is not just about a joy that will come in the future. Having something to look forward to changes the present too. Involving hope and imagination, looking forward to as a form of anticipation can link one with one's future self and open up new possibilities.

While filming and talking with Lara about how she used her smartphone, she enthusiastically anticipated her and her husband Dominic's transformation into home blackberry whiskey makers, a transformation facilitated by her phone.

Roxana: So when would you normally use it [the phone]?
Lara: It's just around all the time. We saw yesterday loads of blackberries around in one of the fields—absolute loads and loads of blackberries. And Dominic's sister was talking about making blackberry whiskey. So the first thing I did was pick it up, and then go in, and look for blackberry whiskey. So I just go into "Search," type in "blackberry whiskey," and it tells me how to make blackberry whiskey. And then I just put it back down. So it's around.
Roxana: So did you check it when you were in the field?
Lara: No, when we came home. When we were around we had this conversation that we had a chat with his sister about making blackberry whiskey. So I said, "right, let's have a look." Oh, and I've also texted her and said, "Can you send me your recipe?" So she then emailed it to me. So I was then able to check my emails [by phone] and there was the blackberry recipe!(transcription from video recording with Lara)

In this "polymedious" (Madianou and Miller 2013) way of using her phone, what Lara was dwelling on was *the possibility* of blackberry whiskey. This was not even a hypothesis one day before. But with loads of blackberry hedges being discovered in the field, and with the easiness of a variety of recipes arriving on her phone from different sources (from Google and from her sister-in-law), making blackberry whiskey by themselves—which is something Lara and Dominic had never done before—became *possible*. After finding the berry hedges by chance, and the whiskey recipes thanks to her smartphone literacy, Lara was in a position where she could imagine herself as a future blackberry whiskey maker. This imagined potentiality, which made her feel enthusiastic, represented one way in which Lara "knew of" her future self: she was imaginatively absorbed with the worldview of her future self, a home-whiskey-maker Lara. The anticipation of this transformation engendered agency.

In his book on modernity and globalization, Arjun Appadurai outlines that "imagination is today a staging ground for action, and not only for escape" (Appadurai 1996: 7). I incorporate his view in the interpretation of the examples of anticipatory actions that I have described so far. The reason why life with something to look forward to is different from life with nothing to look forward to is not because it provides a route to imaginatively escape the struggles or non-eventfulness of the everyday, but because it promises a future state of agentic engagement with the world. When attending the quiz night at the community centre, one performs a series of actions together with others: they work in groups, have a laugh, and maybe give some good answers for which they would be acknowledged. Similarly, Christmas is the most celebrated opportunity for family members to exchange gifts and, in doing so, acknowledge and be acknowledged by the others. Anticipating Christmas is anticipating a day of action when one acts on the other family members through the gifts one makes that show one's knowledge of the others and one's expectations and wishes of how the others should be (see also Miller 1998).

Carol Greenhouse (1996) suggests that we should reconsider agency as referring to "people's goals, together with their broader sense of what is possible and of what relevance is about, even if their understanding of relevance excludes their own experiences from the story of the world" (Greenhouse 1996: 234). The agency engendered in acts of anticipation and accessing a knowing of type of information is not about what one does but about what one *could* do. A future state of agentic engagement with the world is made present just by the fact that it is *possible*: people are able to talk about it and imagine it even if they have not lived it yet. In the end, it is not important if the possibility is acted out or not as long as people are able to extract agentic resources just from the promise of the possibility and as long as they do so with all the possibilities that come their way; from an infinite number of possibilities encountered daily, only a few would be acted upon. Lara is enthusiastic today because she discovered that she could make blackberry whiskey if only she wanted to. The presence of an anticipated future agentic engagement with the world is, in itself, a form of agency because it extends people's sense of what is possible.

A few weeks later, when visiting Lara and Dominic again, I asked about how the plan of making blackberry whiskey unfolded. Lara was proud to show me two jars containing the beverage they made. They were planning to use the small jar this year for Christmas and they keep the big one for Lara's older daughter's twenty-first birthday celebration next year, when they will serve it mixed with champagne, accompanying the birthday cake. They are not planning to use the product for ordinary consumption, but they are saving it for the next two important events. Through this planning, they anticipate how their newly acquired ability of whiskey making will be shared, celebrated with, and acknowledged by a considerable number of kin and friends.

THE MOTHER-MULTIPLE AND ANTICIPATION

I propose to go back now to the process of doing the laundry to discuss the connections between anticipation and the Mother-Multiple that are articulated in this activity. For Cynthia, laundry was usually a weekend task that started with her gathering clothing items and towels that the other family members had abandoned on the floor in the living room and bathroom because they considered them in need of being washed. She would then bring the individual laundry baskets from the bedrooms of her three children into her bedroom and, by taking out clothing items one by one, make four piles on the marital bed: whites, blacks, light colours, and bright colours. However, colour was not the only criterion that she uses in organizing the laundry. While discovering what items were inside the laundry baskets, Cynthia was also concerned with the items that were missing and that she anticipated finding because she *knew* they need washing, such as school uniforms and PE kits, which were usually prioritized and washed in the first two loads of the day. In case of missing items, she would ask or look for them in the related bedroom or inside the bags where she knew that they are normally kept. Sorting the laundry was, therefore, not only a process of selection but also of reviewing, of checking if the *right* items are in place. In addition, for every item that she picked, Cynthia instantly knew the specific washing instructions. According to the washing instructions, she subsequently regrouped one laundry pile into two or more different loads. Moreover, by knowing who each item belonged to and what sorts of preferences or allergies her domestic others had, she could plan the loads according to the type of washing liquid she would use. Thus, in sorting the laundry as Mother-Multiple, Cynthia managed and enacted her complex knowledge about three domains: her family members and their preferences, sensibilities, and allergies; the list of clothing that existed and the sub-list of clothes that were used more often or in "public" situations, such as at school or work; and the material texture of each item and its individual washing instructions.

Anticipation was embedded in this process in many ways. When taking out the contents of individual laundry baskets, Cynthia expected to find specific items, such as school uniforms. Finding these items was part of accomplishing a successful laundry process and she would not continue with the next activity—loading the washing machine—before finding the missing items that needed to be prioritized. With this short-term anticipation, there emerged as well a "looking forward to" type of anticipation in the imagination of the following week when all the family members would wear at school and at work the clothes that Cynthia as Mother-Multiple would have washed and prepared for them. Furthermore, every textile appearance made her rethink the piles and anticipate the moment when she would load the items in the washing machine and choose the programme and water temperature. At the same time, every clothing item represented a person; they were given the names of Cynthia's domestic others and were attributed a specific set of preferences and sensibilities. According to the preferences—sometimes contradictory—expressed within different piles, Cynthia anticipated the moment when she would need to pick the type of washing liquid that would serve as the best solution for each pile while not creating irresolvable conflicts inside the piles.

Cynthia said that once she checked the washing instructions of any new clothing item and followed them, she remembered them every time she saw the item emerge from a laundry basket again. She did not like the fact that her brain retained this information because she considered it rather unimportant and not worthy of being remembered. She preferred to occupy her mind with other things rather than a list of all the clothing owned by her family and the related washing instructions. But she could not help it. Once read, the information from the label remained stored in her mind:

> Sadly, I tend to remember once I've checked it once. [laughing] My brain would be so useful if it didn't keep information like that in it. This is why you can study when you're young like you and you don't have to retain all this information. (transcription from video recording with Cynthia)

In the ethnographic vignette at the beginning of this chapter Cynthia referred back to this conversation while showing me the socks of her domestic others and said, "Again, sadly, my brain knows exactly which socks are whose, so I can do that." By stepping into the Mother-Multiple ontological position, Cynthia became unable to oppose *knowing* a set of information that, as an individual, she would not choose to know. When she said that I could study because I was young, Cynthia meant by "young" a state previous to marriage and previous to having children—as my status was during my fieldwork. Once one starts a family, his or her mind does not exclusively belong to him or her anymore because it needs to accommodate entire new levels of information about people other than oneself and about domains other than the academic. This is what stepping into the Mother-Multiple ontological position entails, letting oneself be embodied by one's domestic others and accessing a consistent corpus of knowledge—about the character, habits, needs, preferences, and dislikes of the domestic others—that can feel overwhelming at times. This corpus of knowledge is tacitly employed in actions of anticipation.

Another domain where the epistemology entailed by being the Mother-Multiple makes anticipation possible is in keeping family calendars. This involves knowing the general schedules and habits of the domestic others as well as a knowing of type of knowledge about what they are doing or what point they are at in their day.

CALENDARS AND TEXT MESSAGES AS INSTRUMENTS FOR ANTICIPATION

What made anticipation possible was often scheduling and routine. "I check the calendar all the time, so I know what's coming" was how Brett formulated his relationship with the future that was mediated by his calendar. Talking about tools for scheduling with the people who took part in my research, it emerged that most of them used two main types of calendars: a personal calendar for work, which was usually a computer calendar and a family calendar for activities and appointments that concerned the other family members in various ways, such as evening social activities, parents' evenings, doctors' appointments, birthdays, and children's activities that represented a change in the routine.

The family calendar was normally a paper calendar hanging on a wall in the kitchen, and the family member who usually filled it in and kept track of the entries was the female adult. Mothers were in charge of this collective time from the beginning of the year when they filled in the new calendar with birthdays of kin and friends by copying them from the old calendar. All family members were encouraged to write down in the calendar their planned out-of-home activities that were due to take place during what was considered family time, namely, weekday evenings and weekends. Eva, who was a homemaker and a mother of two, said about the situations when her husband forgot to put his planned evenings out in the calendar, "If it's not in my paper calendar it doesn't happen, that's my golden rule if he wants to go out." It is the spouse who fills the family calendar first who is able to take the evening off. This situation is similar to a model of social organization identified by Burman (1981) in her fieldwork on the Solomon Islands—and discussed by Munn (1992) in her essay on the cultural anthropology of time—in which the "keeper of the calendar" and the descent group they were part of was able to control the temporal dimension of the everyday lives of all the islanders by being able to regulate "the very motion of time" (Burman 1981: 259). Munn (1992) argues that having control over time is a form of political power.

Here, however, keeping the calendar was a way of enacting the Mother-Multiple. It was a form of caregiving oriented towards one's domestic others. The family calendar made it possible for the person temporarily stepping into the Mother-Multiple to anticipate in one glance a whole month or more in the lives of their domestic others, to imagine the future events that were planned to happen to all family members. Keeping track of what the others were doing, as a form of care, often involved reminding them about their calendar entries. This usually happened through the use of digital media devices, in actions such as phone texting or sending emails, which were regarded to be the preferences of the male adults.

Iris, who was an artist working from home in her shed-studio in the garden, and who also talked about herself as being in charge of the house, kept a paper diary as a family calendar in the study on the ground floor where everybody could check it. "Everything else that's been done by telephone or birthdays and things would be put in the diary, so we can all check it; but it's only me, really, who reads it," Iris said.

From this spot in front of the diary, as a base where collective information about the future could to be found, Iris wrote and sent text messages to her domestic others to remind them about the appointments they had on that day. If the events were taking place in the distant future, Iris's husband Steve preferred to be reminded by email:

My phone is my work phone. So I don't have a personal phone, it's that way around. I might put home events on that calendar, at work–things like tonight that I've forgotten completely about [laughing]. But if kids have parents' evening, I say to Iris all the time, "Tell me, email me". (interview Iris and Steve)

Steve had a senior position at a multinational company working constantly with partners in China and Singapore, so he often stayed late at the office or even woke up in the night to check his email and write replies to overcome the time zone differences and keep the ball rolling. As he spent most of his time at work, the way he dealt with any kind of appointment was through his work calendar. For him, Iris mediated between the realms of family life and work by emailing diary entries that would be transformed into work calendar appointments. Steve admitted that he needed someone to do this mediation so he could keep track of his non-work life.[3]

Text messaging was sometimes used when people wanted to attune their individual times and reach a "common time," which was often the time of picking up, for the situations when they picked their children up from school or from after-school clubs. During my interview with Marilyn, she received a text from her younger daughter who was on a school trip to a museum in a nearby city:

So Helena's just texted me to say they haven't left the museum yet, so she's probably thinking she's going to be late. I normally was supposed to pick her up at 4:45, so I'd probably text her at 4:30 to say, "Where are you on the motorway?" so I knew what time to set off. (interview Marilyn)

While in an interview situation with me, Marilyn knew what she would do when we finished. When the text come up on her phone, Marilyn stopped being an interviewee for a moment in order to step into the Mother-Multiple ontological position. She used her imagination in order to "know of" what her daughter was thinking and could anticipate in detail the next series of actions she would perform to reach the state of being attuned to or in coordination with the time her daughter returned from the trip.[4]

When stepping in the Mother-Multiple ontological position in order to send texts as reminders, people accessed and enacted knowledge of the daily schedules of their domestic others. Just before texting, they often imagined at what point in their everyday schedule the others were and choose to send the text at the moment that would best suit the time of the domestic others rather than their own time. This often happened when female adults reminded their partners of something that they needed to do after work, making sure that they sent the text shortly before their partners finished work, when they were in a temporal area where ideas about the after-work life started to come in and mix with the actions of completing the work day. These calculated actions of anticipating the time of the other in order to find the right moment when to send a reminder involved putting oneself in the other's shoes. An interview with Sam offers one such example:

Roxana: When would you send the text message to remind him? Is it when you remember yourself?

Sam: Yeah, either when I remember—I think "Gosh, I got no milk," or something. Or sometimes I just think I'd better just remind him in case he has forgotten. 'Cause he might say, "I'll get some chicken food on the way home," and then I text him about an hour before he finishes just to say, "Don't forget the chicken food" [laughing].

Roxana: So you'll text him just before his finishing work?

Sam: Yes, an hour, a half an hour. Yes, exactly. If I told him in the morning he might have forgotten 'til lunchtime. (interview Sam)

Knowing someone so well that one could anticipate what he or she is doing when away is something to be proud of. It shows that the relationship between the two is strong and that they know the right ways to reach each other at every moment. Iris talked about texting her husband when he was doing the shopping if she realized that she needed a specific item.

Iris: Steve does the food shopping and I'll text him if I forget something. Because normally he would have his iPod on, so he won't hear me if I ring, so I try and text and see if he'd actually look at his phone. He listens to music when he's doing the shopping, when he's walking the dog, when he's in the house. It's constant music in his ears.

Roxana: And if you text him then the phone would vibrate in his pocket?

Iris: Yeah, it should do! Sometimes he's really good at not hearing it. When he changed it to a Blackberry, the Blackberry wasn't as loud as the other phone, or because it's in a case it doesn't vibrate as much. So he's a little bit better at ignoring that [laughing]. (interview Iris)

Later that evening, when Steve arrived from work, he joined us for the interview. I took him through the same list of questions while Iris informed him about what she already told me.

Iris: And I was saying that if we've gone to shopping as well we might text each other to say, "Remember to get so and so."

Steve: Yeah, even if I don't hear it on my phone because I'm with my headphones on.

Iris [laughing]: That's what I've said! That's exactly what I've said!

Steve: But I always look at my phone before I've checked out, just in case. (interview Iris and Steve)

In this example, there was reciprocity in anticipating each other's actions and in enacting the Mother-Multiple. By incorporating her husband's habits of listening to music, Iris knew that it was better to text him instead of calling for her message to reach its destination. By incorporating his wife's habit of sending him a text when he was doing the shopping, Steve knew to check his phone before heading to the cashier. The reciprocal anticipation of each other's actions made their communication successful and showed a mutual deployment of the Mother-Multiple: a mutual interest with knowing each other very well and finding the way to act that would best suit the other. The success did not only lie in the right item being bought from the supermarket, but in the fact that their relationship was being confirmed and reiterated. Iris jumped with joy and blushed when Steve confirmed the story that she told me before he arrived home. It was suddenly as if they were not taking part in an interview about digital media and time anymore but in a show that tested couples' knowledge of each other.

When people converse through text messages, what happens is not just an exchange of words but also of thoughts. During an interview with Vic and Gail, Vic mentioned that when their daughter, who was in her early teens, went out, he was the one who would normally converse through texts with her from home:

Roxana: So is it just to know where she is?
Vic: It's knowing where she is, it's knowing how she is, it's know-ing what she's doing. (interview Vic and Gail)

Such comprehensive knowledge about what and how his daughter was doing in that very moment could not come just from the content of a text. In the action of conversing through texts, Vic and his daughter would be thinking about each other while typing the messages and while waiting for the reply. In this way, there was an invisible connection that was estab-lished between them and that was at least as important as the words that were exchanged. As he waited for and received the texts, Vic stepped into the Mother-Multiple ontological position and "knew of" how his daugh-ter was by imagining her in the act of texting.

A Form of Monistic Ethical Imagination

The material discussed in this chapter shows that a specific form of "ethical imagination" (Moore 2011) is expressed when people enact the Mother-Multiple. The way in which this ethical imagination operates can be defined as monistic because it does not make an analytical distinction between a person and his or her domestic others. Rather, it expresses the perspective of a nodal point where all the relationships between fam-ily members intersect. The ethical imagination of the Mother-Multiple expressed a common good, a goal similar to what Christensen and Røpke (2010) call the practice of "holding things together": the goal of keep-ing the boat floating in a direction and at a speed that no domestic other would strongly disagree with. In his work on grocery shopping in North London, Miller (1998) finds that people often carry out the act of shopping on behalf of the household, following the normative ethos of thrift, and understanding the general notion of household as a means of transcendence.

However, while the Mother-Multiple provides a means of "transcending" individuality, the focus of the ethical imagination that emerges with this ontological position is not necessarily limited by the domain of domesticity.

In her last recording for the Five Cups of Tea activity, Elaine was preparing for Christmas. She recounted that the previous night she and her children started decorating the Christmas tree. They did not, however, get too far because they realized that nearly half of the light bulbs of their Christmas lights were not working. She said that the kids were getting really excited about the winter celebration, and her son Tom wore his Santa hat to school that day. For them, Elaine would like to finish the process of Christmas decoration soon. She said that every year they took part in a community Christmas window competition that took place on Christmas Eve and was followed by singing carols at the local hospice. This year her husband built a Lego model of the town centre—containing the town's iconic tower, a few houses, and a suspended Santa's sledge—that they would display in their bay window with the background of the Christmas tree. Talking about her plans for the day, Elaine mentioned that she was initially planning to go shopping for an outfit that she needed for an evening out with her friends. However, she was not in the mood for shopping anymore and would rather stay in to sort out the light bulbs for the Christmas lights so that when the kids returned from school they would only have to add the last touches to finish off the Christmas decoration of the tree and window. After talking about her plans for the day, it was time for Elaine to respond to the question that accompanied the last of the Five Cups of Tea, namely, how would you like the world to change:

> I think the thing that cries out to me about the world–and I think it's having children that makes you realize... You bring up your children to share and act fairly, and you paint this beautiful rosy picture of how life should be. And you teach them all the manners and to be courteous–which obviously is the right thing to do. But, especially with Becky, she's eleven, and she's starting to see that there is no fairness in life. And that is life, it's not very fair. You know, there's always people, like, say, business people, that will just kill you ever for their own fat profits. It's all very political. And it's an underlying not very nice world out there. And as the children grow up, they would obviously realize that, even though you teach them the right values. So, I think this is how I would like to change the world: I would like to get rid of greed. That's the enemy. People are too greedy, and they just think

about themselves. So that's what I would change, and then I think the world would be a much nicer place.

Here, Elaine would like to extend a monistic ethical imagination to the entire world: she said she would like to get rid of greed so people would think about the others and not only about themselves. It is greed—a characteristic of some people—that made the world not fair for Elaine. If we would all be able to leave our personal greediness behind, the world would be a much better place.

Elaine shifted her plans for the day from an individualistic preoccupation—going shopping for a new outfit—to a task that responded to an actual unexpected domestic situation and to a set of newly-arisen needs of her domestic others: sorting out the Christmas lights so that she and her family could continue preparing the Christmas tree and window. The form of Mother-Multiple monistic ethical imagination that she employed when doing this shift differed from the ethical imagination of other people, such as bankers and businessmen who are only interested in their own profit, and Elaine knew that. The way she approached this difference was not by assuming a private vs. public dualism, but by wishing that the ethical imagination of the Mother-Multiple would one day be employed by all people in relation to all the others, not just the domestic others.

Getting rid of greed and becoming one another's Mother-Multiple are also ways of moving towards a sustainable future, where global justice would replace the current phenomenon of overconsumption in the Northern hemisphere. By looking forward to this future, one can be empowered to take small but concrete steps towards it every day.

Notes

1. The philosopher Peta Bowden argues that "mothering frequently carries the full weight of ideological constructions of caring. The very nature of caring seems to be produced in the connection between the apparently ultimate vulnerability of early childhood and the potentially perfect responsiveness of mothers" (1997: 21).
2. Which, nevertheless, is not a solution to fuel poverty. For a set of insightful ideas on how to fix fuel poverty in the UK see Boardman (2009).
3. Using a social practice theoretical framework, Christensen and Røpke (2010) identify the practice of what they call "holding things together" as a form of everyday planning and coordination for Danish families.

4. From a different theoretical approach, the sociologist Rich Ling (2004) would regard this situation as an example of "microcoordination" that is made possible through the use of mobile phones.

REFERENCES

Adam, Barbara. 1995. *Timewatch: The Social Analysis of Time*. London: Polity Press.

Appadurai, Arjun. 1996. *Modernity at Large: Cultural Dimensions of Globalization*. Minneapolis, MN: University of Minnesota Press.

Boardman, Brenda. 2009. *Fixing Fuel Poverty: Challenges and Solutions*. London: Routledge.

Bowden, Peta. 1997. *Caring: Gender-Sensitive Ethics*. London: Routledge.

Burman, Rickie. 1981. "Time and Socioeconomic Change on Simbo, Solomon Islands." *Man* 16 (2): 251–268.

Christensen, Toke H., and Inge Røpke. 2010. "Can Practice Theory Inspire Studies of ICTs in Everyday Life?" In *Theorising Media and Practice*, edited by Brigit Brauchler and John Postill, 233–256. Oxford: Berghahn Books.

Cohen, Anthony Paul. 1987. *Whalsay: Symbol, Segment, and Boundary in a Shetland Island Community*. Manchester: Manchester University Press.

Cook, Joanna. 2010. "Ascetic Practice and Participant Observation, Or, the Gift of Doubt in Field Experience." In *Emotions in the Field: The Psychology and Anthropology of Fieldwork Experience*, edited by James Davies and Dimitrina Spencer, 239–266. Stanford: Stanford University Press.

Edwards, Jeanette. 2000. *Born and Bred: Idioms of Kinship and New Reproductive Technologies in England*. Oxford: Oxford University Press.

Ehn, Billy, and Orvar Löfgren. 2010. *The Secret World of Doing Nothing*. Berkeley, CA: University of California Press.

Gabb, Jaqui. 2010. *Researching Intimacy in Families*. New York: Palgrave Macmillan.

Giddens, Anthony. 1991. *Modernity and Self-Identity: Self and Society in the Late Modern Age*. Cambridge: Polity Press.

Grasseni, Cristina. 2007. "Communities of Practice and Forms of Life: Towards a Rehabilitation of Vision?" In *Ways of Knowing*, edited by Mark Harris, 201–223. Oxford: Berghahn Books.

Greenhouse, Carol J. 1996. *A Moment's Notice: Time Politics Across Cultures*. New York: Cornell University Press.

Henare, Amiria, Martin Holbraad, and Sari Wastell. 2007. "Introduction: Thinking through Things." In *Thinking Through Things: Theorising Artefacts Ethnographically*, edited by Amiria Henare, Martin Holbraad, and Sari Wastell, 1–31. London: Routledge.

Irving, Andrew. 2011. "Strange Distance: Towards an Anthropology of Interior Dialogue." *Medical Anthropology Quarterly* 25 (1): 22–44.

Kabat-Zinn, Jon. 1994. *Wherever You Go, There You Are: Mindfulness Meditation in Everyday Life*. New York: Hyperion Books.

Ling, Rich. 2004. *The Mobile Connection: The Cell Phone's Impact on Society*. San Francisco, CA: Elsevier.

Luhrmann, T.M. 1989. *Persuasions of the Witch's Craft: Ritual Magic in Contemporary England*. Oxford: Basil Blackwell.

Luhrmann, T. M. 2002. "Dissociation, Social Technology and the Spiritual Domain." In *British Subjects: An Anthropology of Britain*, 121–138. Oxford: Berg.

Madianou, Mirca, and Daniel Miller. 2013. "Polymedia: Towards a New Theory of Digital Media in Interpersonal Communication." *International Journal of Cultural Studies* 16 (2): 169–187.

Miller, Daniel. 1998. *A Theory of Shopping*. Cambridge: Polity Press.

Mol, Annemarie. 2002. *The Body Multiple: Ontology in Medical Practice*. Durham, NC: Duke University Press.

Moore, Henrietta L. 2011. *Still Life: Hopes, Desires and Satisfactions*. Cambridge: Polity Press.

Munn, Nancy D. 1992. "The Cultural Anthropology of Time: A Critical Essay." *Annual Review of Anthropology* 21 Jstor: 93–123.

Oakley, Ann. 1979. *From Here to Maternity: Becoming a Mother*. Harmondsworth, Middlesex, UK: Penguin Books.

Rapport, Nigel. 1994. *The Prose and the Passion: Anthropology, Literature, and the Writing of E.M. Forster*. Manchester: Manchester University Press.

Reed, Adam. 2011. *Literature and Agency in English Fiction Reading: A Study of the Henry Williamson Society*. Toronto: University of Toronto Press.

Ryle, Gilbert. 1949. *The Concept of Mind*. Chicago: University of Chicago Press.

Simpson, Bob. 1998. *Changing Families: An Ethnographic Approach to Divorce and Separation*. Oxford: Berg.

Strathern, Marilyn. 1988. *The Gender of the Gift*. Berkeley: University of California Press.

Strathern, Marilyn. 1992. *After Nature: English Kinship in the Late Twentieth Century*. Cambridge: Cambridge University Press.

Ungerleider, Steven. 1996. *Mental Training for Peak Performance: Top Athletes Reveal the Mind Exercises They Use to Excel*. Emmaus, PA: Rodale Books.

Williams, Mark, and Danny Penman. 2011. *Mindfulness: A Practical Guide to Finding Peace in a Frantic World*. London: Piatkus.

"Family Time" and Domestic Sociality: Forms of Togetherness and Independence with Digital Media

The concepts of "family time," "quality time," or "time together" emerged during an early stage of my fieldwork when they were described as some of the most important aspects of domesticity. During the first set of semi-structured interviews that I conducted, the people who took part in my research emphasized the idea that home feels "most homely when everybody is at home," as Cynthia put it. Unlike spontaneity and anticipation, which can be employed in one-to-one interactions or in solitude, family time requires all family members to be at home and aware of one another's presence.

In linear temporal terms, family time is generally defined as taking place in the evenings and during weekends. In experiential terms, it is a time of relaxation and celebration as well as digital, material, and alimentary abundance. In relation to domestic rhythms, it can be seen as a set routine that tends to take place every evening even if the activities that make up family time vary. By using the time modes of spontaneity and anticipation that were conceptualized in the previous chapters, family time can be regarded as an anticipated time of spontaneous indulgence. Family time can be anticipated, as Christmas is, because it is a routine expected to happen every evening; its anticipation accompanies one throughout the working day, promising a reward at the end of it. In addition, what family members do during family time varies daily, and in this sense it is spontaneous: people cannot predict what will capture their interest the next evening, what

© The Editor(s) (if applicable) and The Author(s) 2016   139
R. Moroşanu, *An Ethnography of Household Energy Demand in the UK*, DOI 10.1057/978-1-137-59341-2_7

will be the topic of the next friendly argument, or who will win it. The unpredictable character of the evening, as opposed to the general predictability of the working day, is important in defining evening family time as a special experience. The people who took part in my research often emphasized the variety of technologies and other media that they could choose from or combine in new and creative ways every evening. The results of unexpected findings and new assemblages would often bring new experiences for them, such as learning how to play a new game on one's digital device, hearing news from a friend who lives across the ocean, having an interesting chat with the other family members, trying a new type of snack, or discovering a new way of streaming music or video content.

Nevertheless, family time can be regarded as a normative social construction. In Gillis's (1996) terms, family time is an "ideologically constituted form of prescription" (1996: 17) that has the power to convince family members that togetherness is desirable and pleasant even if their everyday experiences of it might contradict this belief. The idea of family time cannot be disentangled from the ways in which the concept of family or, more specifically, middle-class family (Strathern 1992) has been constituted and expressed as an ideological unit in Euro-American contexts (Collier et al. 1982). In these contexts, it is a concept that "imposes mythical homogeneity on the diverse means by which people organize their intimate relationships" (Stacey 1990: 269) and that is extensively politicized. In the case of the UK, Strathern (1992) looks at how the concept of middle-class family was redefined in the Thatcher government, while Silva and Smart (1999) analyse the appeal to family as a "pillar of supposed stability" (1999: 2–3) in the political rhetoric of former prime minister Tony Blair. These two analyses show that the concept of family can be approached as a tool of governance from both sides of the political spectrum in the UK through the assumption that most families operate in similar ways (Finch 1997).

In relation to social anthropological literature on English kinship, I suggest that family time is a concept used to measure the quality of home life in "family-style lifestyle" (Strathern 1992) settings. Moreover, making family time, an active process that involves ordinary actions and routines, can be regarded as an everyday practice of creating relatedness (Carsten 2000a). While my approach pays attention to people's agency in making and maintaining this time of togetherness, I acknowledge the coexistence of a cultural imaginary of the ideal family operating though social prescriptions that link having a family with having family time. I approach the tensions between a social constructivist perspective on family and family time, and a focus on people's agency in developing a process of relationality (Gabb 2010), of constituting trust reflexively through everyday acts of care and intimacy (Williams 2004) as a way of "doing family" (Morgan 1996), through the lens provided by Strathern (1992) in her discussion of the relationships between convention and choice in English middle-class kinship. Thus, I argue that *by* keeping family time and by emphasizing its importance over other dimensions and qualities of the home, people "do" convention and show their capacity for morality. However, *what* people do during family time is a matter of choice that reflects individuality and diversity, which are regarded as twin concepts and the first two facts of English kinship (1992). Having and talking about having family time shows my research participants' "normality," their following of a social norm. Talking about the creativity and eccentricity of the assemblages that they develop during this time shows their difference and their capacity to transgress this social norm while keeping it. These transgressions are ordinary events, such as having a desert dinner instead of a "proper" hot meal[1] or keeping the curtains closed for a whole day. However, the existence—or even their necessity—of these transgressions shows both the rigidity of social norms regarding family life and people's awareness of this rigidity and their agency in challenging it.

English Middle-class Kinship

Contemporary studies of kinship in Britain have focused on topics as varied as new reproductive technologies (Edwards 2004, 2008; Edwards et al. 1993; Edwards and Strathern 2000); place-making, identity, and belonging in relation to kinship thinking (Edwards 2000; Strathern 1981); adoption reunions (Carsten 2000b, 2004); and divorce and separation (Simpson 1997, 1998). The recent theoretical advances in this field, such as Carsten's (2000a) development of the concept of "relatedness" as an alternative to kinship that conveys "a move away from a pre-given analytic opposition between the biological and the social" (2000a: 4) that framed the first wave of anthropological studies of kinship, are based on the initial denouncement of this opposition (Schneider 1968, 1984; Strathern 1992), which influenced all the subsequent approaches to kinship. I will summarize this critique as part of Strathern's (1992) seminal analysis of English middle-class kinship as a system of thought, which represents the theoretical backbone for this chapter.

In her work on kinship thinking in English middle-class contexts, Marilyn Strathern (1992) uses Victorian middle-class kinship constructs and British anthropological kinship theory—which was one of the main theoretical directions that British social anthropologists were concerned with in the first half of the twentieth century—as "mutual perspectives on each other's modernisms" (1992: 8). Her endeavour was inspired by the work of the American anthropologist David Schneider. In his analysis of American kinship (1968), Schneider argues that this type of kinship system is essentially defined by the two orders of nature and law—or sexual reproduction and marriage, respectively—where the role of the "natural" or "biological" is substantial. In a further publication, Schneider (1984) demonstrates that the biological predicament—sexual reproduction—is a Western ethnocentric assumption that has been brought into the anthropological analysis of non-Western kinship systems. Strathern takes this idea further, arguing that "the unthinking manner in which generations of anthropologists have taken kinship to be the social or cultural construction of natural facts" (Strathern 1992: 45) was the pivotal stone in the construction of kinship as a domain of anthropological inquiry. Moreover, Strathern remarks that if kinship used to be a domain that connected the realms of nature and culture in the Western world, this situation began to change in late-twentieth-century Britain in a post-Thatcherite neoliberal socio-political climate with the appearance of new reproductive technologies. Thus, nature can no longer be regarded as an underlying predicament for kinship.

Strathern argues that in post-Thatcherite Britain, "family" is a specific type of lifestyle that one can opt for: "The family as a natural consociation vanishes in the promotion of family-living as an experience" (1992: 147). Furthermore, if the emphasis is placed upon the experience that a family-style lifestyle could provide, then "[t]he family as a set of kin relationships disappears in the idea that the quality of home life has an independent measure" (1992: 149). I argue that one way of expressing the experience of family-style living, and of measuring the quality of home life, was enacted by the people who took part in my research through the use of the concept of, and by actively making, family time. I will come back to and develop this point in the next section.

In this context, where family is seen as a type of lifestyle that anybody can embrace if only one chooses to, Strathern argues that substantial emphasis is placed upon choice. Here, "choice" comes to mean "individuality," while all the other possible ways of defining an individual are minimized. Strathern calls this type of individualism "prescriptive individualism":

In the late twentieth century it is possible to think that morality is a question of choice. Prescriptive individualism: choice requires no external regulation. As a consequence, the individual is judged by no measure outside itself. It is not to be related to either nature or society (*vice* national culture). It is not analogous to anything (1992: 152; original italics).

Furthermore, Strathern identifies a complex relationship between choice and convention, which is a point that I build upon in this chapter. Strathern argues that in "the modern epoch" (1992: 154), convention was a concept that regulated social life considered to be "the cultural counterpart of natural law" and as "embodying the order necessary for sustaining a complex (and civilised) life" (1992: 157). Thus, convention defined general social norms of good conduct. However, in a post-Thatcherite Britain, where individual persons and families take the place of society, convention is no longer understood as external. Rather, it takes an expressive function and is "internalized as personal style" (idem: 158). When convention is not external to the individual but internal as choice used to be, one could opt to display convention as right acting. "But the consequence is that," Strathern argues, "the individual person comes to contain within him or herself the knowledge for right acting, and thus becomes his or her own source of morality" (1992: 157). Thus, in the Thatcherite approach, the fact that "the people will know what is right is taken for granted" (1992: 159) and an analogy between the individual and the family is created.

Following this analogy, if the individual person is his or her own reference point, the middle-class family is also assumed to contain within itself the knowledge for right acting and can be seen as its own reference point. Families appear, therefore, as distinct universes regulated by internal, self-imposed norms and are judged by an internal measure.[2] This embedded set of criteria that white middle-class nuclear heterosexual families follow in order to legitimize and to measure themselves takes into consideration the quality of home life and can often provide a personal internal scale, for example, how our life as a family was before and how it is after we had the third child, how it was before we built the extension and how it is now, or how we experience our evenings together in the living room now that we have smartphones and tablets. The people who took part in my research compared their experience of home life to itself, to how it used to be before a specific domestic event or a new purchase occurred, and not to other people's experiences of home life or to a generally accepted idea of how home life "should be."[3]

If people "do" convention by choosing a family-style lifestyle, and this deliberate act is enough to show their capacity for morality, then the way they carry out their home life can be as "unconventional" (e.g., having a dessert tea instead of a "proper" hot meal or spending a whole day with the curtains closed) as they wish. In my fieldwork, the ways in which different types of everyday "eccentricities" were performed and displayed marked the individuality and diversity of the families. Vic and Gail described one interesting example of such everyday eccentricity when I asked them whether they would ever use a moment in a TV programme to coordinate activities inside the home.

Vic:	There is one I can think of.
Gail:	Oh, for goodness sake!
Vic:	We have a family joke. There is one particular programme called New Tricks. It's not a particularly good programme, but it's got this singing tune. Everybody has to sit down when the theme tune's on and go like this [he shakes his arms up and down, singing along "It's all right"]. So, you've got all of them singing here around. So, that's the only one I can think of. It's a family joke, really, more than anything.
Gail:	Everybody comes in here [the living room] and does it. Even my mother did it.
Vic:	Even to the extent of stopping what they're doing. They come here, and do it, and then go out when the programme's on. Gail's come in from the kitchen, sat down, done it, and gone back in the kitchen again. You have to do it here, in the room, either on the sofa or in one of these two chairs. It used to be Friday nights; now it's Mondays I think. (interview Vic and Gail)

This example of a family joke is a type of "family practice" (Morgan 1996) through which people create their own family as "special," as "different," as a distinct universe. As I will show, the kind of families that the participants in my research were doing expressed individuality and diversity while, in Strathern's (1992) words, also being their own reference point.

FAMILY TIME: MAKING THE EVERYDAY EVENTFUL

The concept of family time was one of the ways in which the people who took part in my research expressed the experience of family-style living, and measured the quality of their home life. This idea first emerged as a response to one of my interview questions about the possibility of having various degrees of feeling homely in one's home. The general response was that "it's most homely when everybody is at home," as Cynthia put it. This is how Sam expressed this idea, by linking it with her husband's work shifts:

> I suppose it's a little bit more homely in the evening when everybody is back from school and work. Weekends are always quite homely because we're all around—we're in and out. Saturdays afternoons and Sundays are kind of family time as much as we can with Peter's job, obviously. (interview Sam)

People sometimes emphasized the importance of family time over the spatial and sensorial qualities of their homes. For example, they said that they did not mind what their living room looked or felt like as long as they could spend time together there as a family. They also mentioned that they preferred to use their free time by spending it together with the other family members rather than by cleaning the house.[4]

From a social constructivist perspective, Gillis (1996) argues that family time, as a prescription, has nothing to do with the ways in which togetherness is experienced in everyday life. As he puts it, "family times tend to be anxiously anticipated and fondly remembered, but, as events, they are often experienced as stressful and frustrating, because it is when families are together physically they are furthest apart in terms of their generation and gender assignments" (1996: 16). He argues that the strength of the concept of family time comes from its modern definition of ritualized time that can coexist with other forms of time, such as the linear time of the capitalist market. Rituals are argued to have the capacity to suggest stability, and, in Giddens's (1981) terms, they have an important emotional dimension as a source of "ontological security" and also provide a sense of history and tradition. In the case of family life, as Gillis argues, "ritual provides not only those moments when families are actually with one another, but, more important, when they imagine themselves as families" (Gillis 1996: 15). In other words, maintaining and enacting family time as a ritual means maintaining (the imagination of) the family.

For Gillis (1996) family time is, therefore, a cyclical ritual that coexists with the linear time of the capitalist market, which promotes a logic of progress. However, here I regard family time as one of the time modes of domesticity that constitutes itself as a counterdiscourse to linear time by making the present future. Evening family time is experienced as continuous, as I will later discuss in more detail; it is not punctual, but a sequence that people actively extend using various strategies, such as closing the curtains early, having repetitive snacks and drinks, or making themselves stay up late. Family time incorporates the short-term future—once started, nothing will stop it before bedtime—and it usually ignores the long-term future. Family time is a form of seizing the day and not worrying about tomorrow. If it was to be regarded as a grammatical tense, family time could be seen as the present perfect continuous: it is a present that goes on and nobody really knows when it will transform into the future. However, the night provides a natural stopping point to family time. To add one more layer to the description of family time, one could correlate it with what the philosopher William James called the "specious present" (1890). For James, the concept of specious present suggests that the moment of *now* that we experience as present "is not punctuate, but rather includes a small but extended interval of time" (Andersen and Grush 2009: 278). James (1890) argues that the specious present is most visible in the ways in which people perceive movement: as movement happens during an interval and not just in the moment of *now*, the fact that human beings can perceive movement means that we occupy the present as an interval rather than as a moment. One of the ways in which the present as an interval is particularly acknowledged and celebrated during family time is through a specific form of domestic sociality that I propose to call "physical togetherness and digital independence," which will be discussed later.

The form of "ordinary agency" that family time engages is related to challenging the uniformity of a dominant social ideology of (middle-class) family, through the enactment of everyday eccentricities and family jokes as well as unconventional practices, habits, and actions. Family time can thus be seen as an established time when families emerge as distinct universes that express diversity and individuality and when they constitute themselves as different from other families and the uniform ideology of (middle-class) family. The form of agency engendered by the time mode of family time is, therefore, related to creativity and the work of creating difference, which constitutes a tacit and ordinary form of resistance to social prescriptions.

ELUSIVE PRESENCES AND DIGITAL MEDIA

In a publication drawing from a research project carried out in the North of England about children's understandings of time, Christensen (2002) outlines that, for children, the qualities of time are different and much more varied from what is usually called "quality time." Thus, instead of quality time as a nominated block of time when all the family members do activities together, children have other views and preferences for their time at home, such as the situation of being in separate rooms and doing different individual activities while knowing that the parent is available somewhere in the house and could be called in case they are needed. This brings out the idea that family time is not necessarily time together but time when everybody is at and around home. In Sam's words, quoted in the last section, "weekends are always quite homely because we're all around—we're in and out."

This state of flexible togetherness is achieved when people are aware of what the other family members are doing, generally from sounds—footsteps, music, or other forms of activities—or the smells of cooking, but they are not directly taking part in those activities. However, the sense of shared time and the feeling of homeliness emerge from such situations as much as from more accentuated forms of family time. While making a Tactile Time collage[5] about her son's usage of the Wii (Fig. 7.1), which was located in the spare bedroom, Sam chose a loose transparent fabric to express how that time feels for her, and she explained her choice by saying:

> Because he's here, but he's not here: you can hear him running, jumping about upstairs, but you can't see him or anything like that. (interview Sam)

Thus, Sam was aware of her son's presence in the house, and she could tell what he was doing from the sounds he was making: she could imagine him performing the actions required in his video gaming, such as jumping and running. Even if he was not in the room and she could not see him—he was not physically co-present—Sam still considered her son to be "here." His presence was elusive; he was part of the world, part of the universe of the house as the upstairs bedrooms were, and Sam did not need to see him or the rooms to accept that her experience of home was framed by these elusive presences.

Another way of conceptualizing forms of contingent relatedness and everyday intimacy in families is through the notion of "quality moments"

Fig. 7.1   Detail from Sam's Tactile Time collage showing her son's usage of the Wii

developed by Kremer-Sadlik and Paugh (2007). In their research on everyday life with American families, Kremer-Sadlik and Paugh propose the concept of quality moments to complement the notion of "quality time." Quality moments are "spontaneous, unstructured, everyday moments of shared social interaction between family members" (2007: 288) that can appear ad-hoc during household chores or while waiting. The researchers were surprised by the frequency of these interactions during activities oriented towards other goals. They emphasize that quality moments are closer to the experience of everyday life, while quality time stands like a normative ideal model that is rarely fully achieved in practice.

In some situations, digital media are used inside the home to assert and make manifest the elusive presence of other family members. These interactions can be seen as spontaneous quality moments (2007: 288) when people create and reinforce relatedness—or, in Morgan's (1996) conceptualization, they are *doing* family—through ad hoc creative employments of digital media. Besides the connection between family members, digital devices also connect two essentially different parts of the home in the following examples: the upstairs and the downstairs.

The vast majority of English houses comprise of a ground floor and a first floor.[6] The ground floor is a communal and generally daytime space comprised of the kitchen, living room, and, sometimes, a front room (that used to be called a parlour) functioning as an occasional dining room for guests and that is nowadays often used as an office space or an extension of the living room.[7] The first floor, which consists of individual bedrooms, is generally a night-time realm or the nominated area for voluntary or imposed solitude. The ground floor is the space for receiving visitors, while the first floor is considered more private and is usually reserved for the members of the nuclear family. Children's bedrooms, however, can be open to visitors, such as their friends who might visit for a sleepover on the weekend. The main double bedroom, occupied by the cohabiting adult couple, is usually the most private room in the house.[8] During my fieldwork visits, I was always invited into the kitchen, conservatory, or living room—social spaces equipped with tables that could hold cups of tea. The only times when I had access to the first floor was to film a video tour of the house or re-enactments of practices, such as laundry and bathroom practices. The mono-functionality of the bedroom as a room for sleep was often emphasized when the adult research participants described their children's bedrooms by saying that the kids "do not spend any time there at all" (in the daytime) because they are normally downstairs. Children tended to play and do their homework downstairs, and they would usually rediscover the bedroom as a personal space in their teens. The ground floor of English houses is, therefore, typically a communal space for cooking and eating, leisure, receiving guests, and for work, while the upstairs represents a space for rest and individual pursuits. While at home and, most often, during the weekends, people navigate between these two spaces as they navigate between collective and individual activities. However, digital media are sometimes used to connect people on different floors and to connect the realms of individuality and commonality. Sam talked about the ways in which her daughter Julie, aged eleven, uses video chat computer software at home:

Sam: She just Skypes me from upstairs sometimes [laughing]. She's got Skype on her iPod. So, I had the odd call on. And she Skypes her friends on her iPod. She Skypes quite a lot, actually.

Roxana: So you were at the computer downstairs?

Sam: Yes, I was. She doesn't do it very often, but at odd times she might find it funny to sit upstairs and Skype rather than get up and come down, whatever. (interview Sam)

Julie's spontaneous use of Skype while she is in her bedroom to contact her mother downstairs is a way of asserting her presence while still keeping it elusive. It is not Julie's body coming downstairs to talk with her mum, just a visual and aural digital representation of Julie that breaks the time and space of individuality to connect with Sam. Skyping keeps Julie both separate and connected. In this example, Julie's employment of Skype is seen as whimsical, part of a realm of improvisation, creativity, and play. However, this is not the only situation when digital devices are used between family members inside the house in Sam's family. She recalled that the only conflict they have over TV programmes is on Saturday evenings when she and the children want to watch an entertainment programme called *The Magicians* while her husband wants to watch a football game that he would have recorded during dinner. Because the recording is stored on the V+ box that they have with the TV in the living room, Sam and the children would have to go upstairs to her and Peter's bedroom ("we are banished to the bedroom") where they have a second TV. They would watch the programme snuggling in bed while Peter is downstairs by himself. In this situation, the communal ground-floor space is experienced by oneself in isolation while the rest of the family is regrouped in a bedroom upstairs. However, Peter would not watch the whole game. He would skip through it, watch it at double speed, finish it in half an hour, and call the rest of the family back downstairs using the internal phone system. This type of exceptional situation, which happens only on Saturdays, was recalled by Sam twice several months apart: during the video tour of her house and during one of my interviews. This shows that the use of the internal phone system to reunite the family downstairs, even if playful, is not "out of the ordinary." It is part of the Saturday evening routine and can be seen, in Morgan's (1996) conceptualization, as a family practice.

Other situations when family members use digital media between themselves inside the home are for various requests that might sound different— and maybe less demanding—when asked through a digital platform than when asked directly or in a face-to-face form of interaction. This might happen at dinnertime when all family members are asked to stop their individual activities to take part in what is usually considered to be an important family ritual.[9] Some research participants recall that media, such as the TV, used to be a motive of discord at dinnertime when children wanted to finish watching their programmes while parents wanted them to sit around the dinner table. Nowadays, when they have the ability to digitally record and store programmes on their TV box, people can prevent these arguments. Cynthia explains that her family would use the record function spontaneously in relation to contingencies or routines that might result in overlapping activities:

> The recording is really good with the children 'cause you're watching a programme, and it's time for the meal, it's time to do something—just record it and do what you were supposed to do. Because, before we had that, there would be lots of arguments—"I just want to watch the end, the last ten minutes." (interview Cynthia)

Here, the capacity to record is similar to an ability to multiply time. Instead of choosing to spend the ten minutes doing just one activity—join the dinner table or watch the end of the programme—people can do both by pursuing the planned activity first and getting back to watch the last ten minutes of the programme afterwards. Cynthia also discovered other ways of using digital media for easing the transition between children's individual activities and dinnertime. She mentioned that she sometimes sends instant chat messages to call her teenage and young adult children for dinner when they are upstairs in their bedrooms:

Cynthia: If I know Kelley is online, and I've got my laptop on, and the dinner's ready, than [I write] "dinner's ready!" Yeah, we use Facebook.

Roxana: When Kelley's in her bedroom and you are downstairs?

Cynthia: Or Lee. But Lee's not often on Facebook though. Nine times out of ten, I just shout "Food!" up the stairs. (interview Cynthia)

By using Facebook, Cynthia displays her knowledge of her children and the possible activities they might be engaged in at that time of the day. As in Julie's case, the connection between upstairs/individuality and downstairs/commonality is realized virtually or aurally through the medium of Cynthia's voice. However, the employment of digital media is seen here as special because it happens in just10 % of the cases. This "out of ordinary" characteristic, together with the quality of digital communication potentially being less intrusive, might produce a better response.

In other situations, the choice of using digital media instead of having face-to-face interactions inside the home is explained as being more convenient, in terms of effort, for the sender of the message. Iris and Steve recalled a situation when Iris sent a text message to their teenage son, who was in his bedroom, to remind him to have a shower:

Steve: Iris was texting Alan the other night to tell him to get in the shower.

Iris [laughing]: I texted him from the sitting room because I couldn't be bothered to get up! I forgot about it. It's ten o'clock at night. I can't be bothered to get up! (interview Iris and Steve)

Here, the possibility of sending the text gives Iris the chance to keep being responsible of her children's evening baths while responding to her own needs as well, such as the need to sit down and relax in the evenings. Iris suggested that "mother time" should finish before ten o'clock in the evening when the children would have already gone to bed, and her choice of sending a text rather than getting up and climbing the stairs to her son's bedroom shows her drawing a boundary between being "totally available" and "partially available" for her children.

These examples show that a specific form of domestic sociality emerges when family members share the overall environment of the house while not being in the same room. This form of sociality, which I propose to call elusive togetherness, is uni-sensorial: it is enough for the existence of the others to be signalled through only one impression—be it aural, olfactive, or visual—in order for the situation to be recognized as sociality by both the producer of the signal and by the recipient. A uni-sensorial signal is enough to trigger one's imagination of what the other is doing and where in the house they are located as we have seen in the example of Sam and her son who was videogaming. The imagination of the other, who is located in a different room of the house, changes the way in which one experiences one's home.

The way in which I understand the concept of sociality follows the approach proposed by Long and Moore, who regard sociality as a "dynamic relational matrix within which subjects are constantly interacting in ways that are co-productive, and continually plastic and malleable" (2013: 4). Long and Moore argue that human sociality should not be reduced to affectivity or relationality; rather, it should be situated by attending to what is specific about human beings in employing "an explicit theory of human subjects" (2013: 3). Some of the human specificities that they identify are the imaginative, motivational, ethical, and representative—referring to attaching meaning and significance to things—capacities that people can enact. In creating an elusive togetherness form of sociality, my research participants made use of representation, such as when sounds signified footsteps; of imagination, such as in the example of Julie imagining her mother downstairs reaching the computer to pick up the Skype call; and made appeal to various motivations, such as the wish to have a nice dinner time that all family members would enjoy, in performing intentional actions, such as using Facebook to tell one's teenage children that the dinner is ready.

EVENING TIME ROUTINES: PHYSICAL TOGETHERNESS AND DIGITAL INDEPENDENCE

In order to describe the forms that family time takes during evenings, besides scan copies of some Tactile Time collages, I will employ in this section another type of material that I have not yet drawn from explicitly in this work: the videos that the participants in my research produced as part of the Evening Times video diary activity. Here, I employ the knowledge gained from these videos by transforming the content of some clips into short narrations. My intent in using these narrations is to illustrate the perspectives that my key participants provided over their family time while family time was happening. In the corpus of this section, the narrations are differentiated from the rest of the text through indent alignment.

Evening times, the times when all the family members are around and when the home feels most "homely," were described as established segments of leisure and relaxation. Often abundant, in terms of the frequency of food and beverage items that were consumed, they marked a conspicuous break from the non-memorable character of daytime work tasks and household chores. They were little celebrations of family, individuality, and of life in general. Following the ways in which this temporal segment was represented in the Tactile Time collages, "evening time" can be defined as starting after dinner, when the primary family TV is switched on and people gather around it, and finishing when the TV is switched off and the last family members to watch it head upstairs to their bedrooms.

In these collages, felt is the fabric that was chosen in all the cases to illustrate how evening times felt. There were three different types of felt of various degrees of softness in the collage kit, the softest—as agreed by participants and by me—was the pink one. Pieces of pink felt appeared in all the representations of evenings and, in some cases, of other moments of the day as well. This choice was explained by expressing feelings of warmth, cosiness, relaxation, and "being settled down." In Cynthia's words, "I've got felt because I'm fed, relaxed, ready for a nice evening. It's softer than this; this is rough, but warm. This is the softest, it's a relaxed fabric" (transcript from video of Tactile Time collage making with Cynthia's family).

Evening times imply the imperative of "indulgence." I situate the idea of indulgence by using the dichotomy of thrift and treat developed by Miller (1998). In his ethnography of grocery shopping in North London, Miller identifies that the most important element in buying provisions for the household—which is understood as an entity in itself that transcends the individual—is thrift, or the experience of saving money. He describes in detail his participants' strategies for finding savers in the shop or by choosing different shops for specific categories of products, and he outlines that for the activity of shopping, thrift can be seen as an end in itself. In this context, the treat is "a slightly transgressive purchase," an "extra extravagance that lies outside the constraints of necessity" (1998:40–41) and is directed to a particular individual, usually the shopper. The treat, which can sometimes mean just eating a grape or two from the fruit stalls in the supermarket, can be seen as a reward to the shopper for carrying out the act of shopping. Miller argues that the treat is "an action which specifically reaffirms the self" (1998: 47) by individualizing the recipient of the treat. In Miller's ethnography, thrift and treat appear as a pair: treat is seen as a (instant) reward for the success of achieving thrift.

However, "indulgence" at evening time is a prolonged state of treat that does not necessarily follow the successful accomplishment of a task or "duty" but is enacted as a way of expressing agency. If daytimes are organized following the thrift and treat logic when people might motivate themselves with treats for carrying out tedious or non-inspiring tasks, evenings follow a totally different order. During the evenings, indulgence is an imperative: the *forms* of indulgence express individual identity and the *enactment* of indulgence shows agency. In my research, indulgence involved food and beverages, digital media and other technologies that were used simultaneously, and a general mood of total relaxation and "carelessness" in relation to one's duties, which could be described through the expression "I can't be bothered" (as Iris is quoted as saying in the previous section).

What I have found to be a general imperative for evening time indulgence is similar to the Swedish phenomenon of Cosy Friday (Brembeck 2012), an established ritual for families with children that consists in gathering together on the sofa in front of the TV and is defined through an abundance of food (usually fast food and snacks) and beverages. Brembeck outlines the importance of the Cosy Friday feeling as providing "a special *experience* of freedom, relaxation, and togetherness" (137; original italics) in opposition to everyday experiences of work and order. Unlike the general evening time indulgence that I encountered in my fieldwork, the Swedish phenomenon is child-centred (2012:137) because it represents a time when no alimentary treats are refused to the children and the family TV choices are generally children's programmes and films.

Below, two Tactile Time collages made by Lara, Dominic, and their twelve-year-old son Ewan visually depict the difference between what can be called the thrift-and-treat logic and evening indulgence. The first collage (Fig. 7.2) shows the family's usage of the computer during the

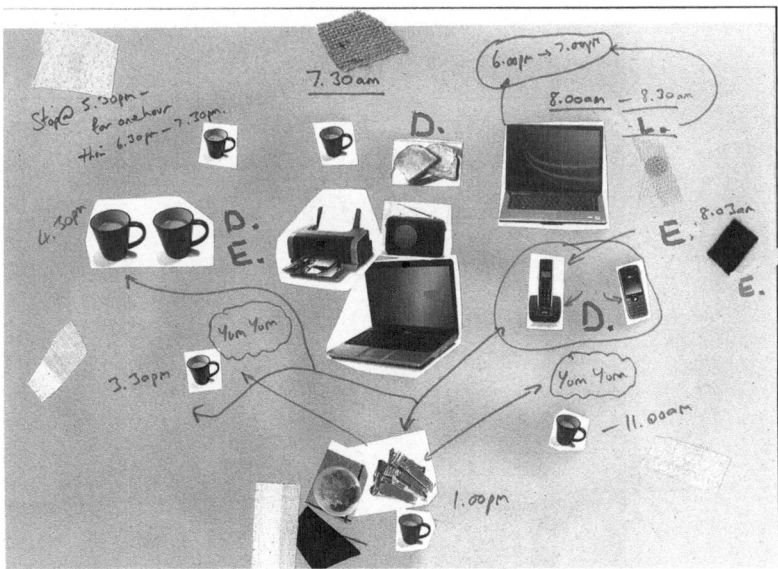

Fig. 7.2   Tactile Time collage made by Lara, Dominic, and Ewan showing their laptop usage

day, together with other technologies and food and drinks. Lara used her laptop in the morning to check her email before going to work. Ewan used it, together with a drink, in the afternoon when he came back from school. Dominic, who worked from home, used it all day. He had a break with a cup of tea and a snack at 11 a.m. and at 3:30 p.m., and at 1 pm he had lunch. In this collage, the drinks and the snacks placed around the computer are visually proportionally distributed during the day, between 7:30 a.m. and 7:30 p.m., with a one-hour dinner break. They mark and break the time that Dominic spent on his own and the series of work tasks he performed, which might also have involved the landline phone, the mobile phone, and the printer as shown. In the second collage (Fig. 7.3), the right half of the picture shows only Ewan's TV usage during two time slots—between 7:30–8:10 a.m. and 4:30–6:00 p.m.—while the left half, which represents the family's evening time, shows an abundance of digital devices, food and beverage, and other media such as books.

Fig. 7.3 Tactile Time collage made by Lara, Dominic, and Ewan showing their TV usage

The living room is dark. The only light that is on is a lamp behind the TV. The camera, held by Dominic, looks at Ewan: he sits on the sofa wearing his pyjamas, with a laptop in his lap and wearing big headphones; he says "hello" in a small voice. Lara sits on the other sofa, her feet under her, using her iPhone. Their two dogs keep on moving in and out of the frame. The TV screen shows a programme about relocating to the countryside. Dominic follows the dogs to the kitchen. In the sink, one can see a baking tray sitting vertically. Dominic stops by the side table where two full cups, one of purple and one of brown hot contents, wait to be picked and transported. The recording stops.

Next, it is a Saturday afternoon, and the living room looks bright. On the sofa: dog toys, big black headphones, cables, and one laptop. Lara sits at the living room table checking her phone. In the kitchen on the wooden table, Dominic's laptop shows a black scree with white code language, rebooting. Something went wrong with his laptop, and he is filming while waiting for the computer to start working? The ginger cat sits in the door, and the dogs approach him with an inquiring glance.

As a time for celebration, abundance, and indulgence, my research participants often liked to be able to extend evening time, to make it feel longer. In Sam's collage below (Fig. 7.4), evening time, which lasts between 7 p.m. and 10 p.m., is represented through a longer piece of fabric, which shows this time as continuous.

A twenty-nine-minutes-long static clip shows Sam, Peter, and their children Julie and Alex sitting on the living room sofa, and watching *The Voice*. The video camera is placed on top of the TV, and the programme can only be guessed from the sound. However, the sound is what Peter mainly experiences from the programme as he is reading a magazine—and later, the printed newsletter of the school or a piece of homework that Alex shows to him—and he very seldom lifts his head to watch the screen for more than a second. At the beginning, Alex can only be heard; he cries, "not again!" or "oh, yeah!" from time to time. Suddenly, he appears holding a miniature pool cue stick. Later, Sam leaves the frame, and the others start chatting about Alex's game of soccer. He brings a card box lid that says "10 sport games in one" to his sister on the sofa. We then hear him asking Sam to play a game with him and her playing the game (she is losing, he is winning). Sam returns with a plate containing a crumpet with chocolate spread for Julie. Then, Alex appears holding a similar plate, and he comes to the camera saying, "Hi, we are watching *The Voice*" and happily showing us the content of his plate. He goes to the sofa and sits down on his mum's

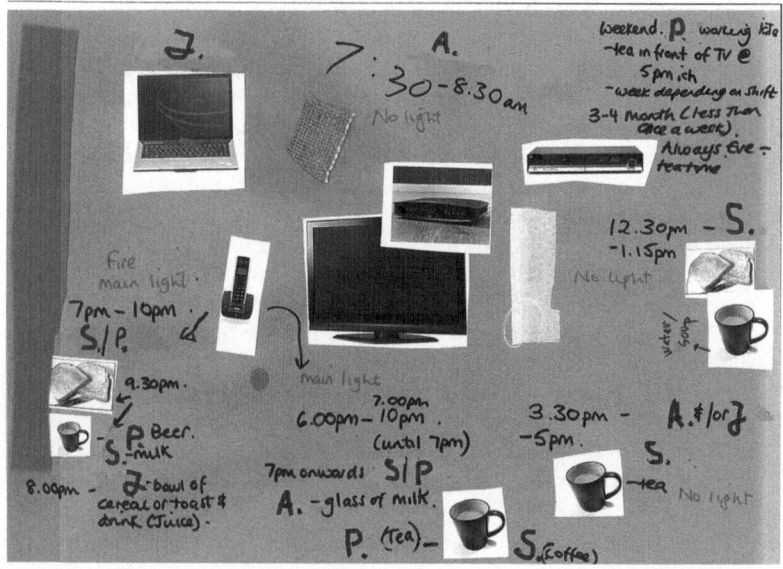

Fig. 7.4   Tactile Time collage made by Sam showing her family's TV usage

lap. The next fifteen minutes pass in this way: Sam sitting between her children, embracing them, and Peter sitting beside them near Julie, switching between his reading material, the TV screen, and responding to what the others say. They all keep their feet on a footstool, and they look very comfortable. The children eat their crumpets, and they are all chatting about the choice of songs and about the contestants' performances.

Other research participants mentioned their strategies of extending evening time. While she and her family were making the collage below (Fig. 7.5), Cynthia explained her entry for 9:30 p.m. that is illustrated by a photo of toast and a dark red taffeta choice of fabric. "I start to feel a bit rough, so I try to make myself better by having some toast. Probably I'm ready for bed, to be honest, but I make myself stay up."

"I thought I'll do my filming in the last few minutes before I go to bed." This is how Cynthia's first recording starts. While she is filming, Cynthia is talking with the camera, explaining the situation like a reporter. She is in the living room together with Jeff, and she explains that he is researching "fam-

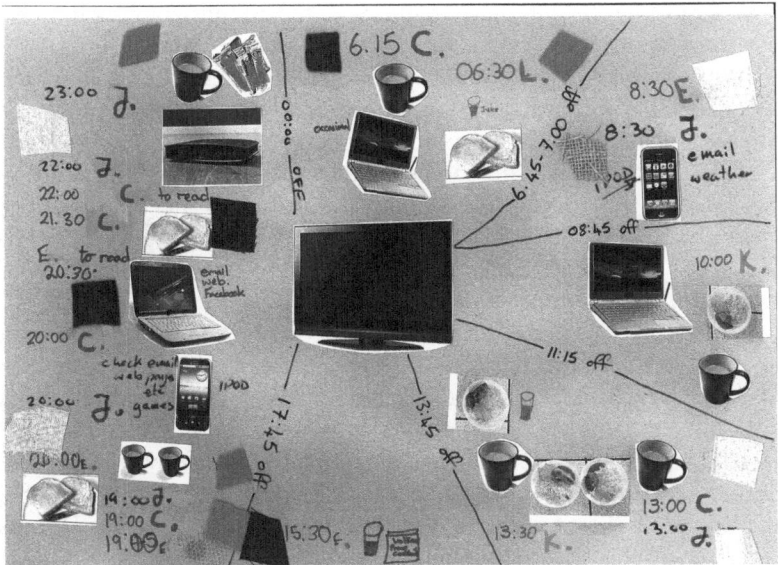

Fig. 7.5 Tactile Time collage made by Cynthia and her family showing their TV usage

ily tree stuff" on his laptop. "The TV is off, but as soon as I sit down, it will go on because we can't both look at family trees at the same time," she continues. She heads towards the window, and she closes the curtains. Next, she visits the kitchen, where her son Lee is doing the washing up while listening to music from his iPod that is plugged in a docking station on top of the fridge. She is amused by the name of the song and by the fact that her son is multitasking: he is preparing a snack for himself—toasted crumpets—while finishing tidying up the kitchen. Later, she goes back to the living room and sits on the sofa by her husband, asking about the new family tree discoveries and watching the weather forecast on TV. The camera moves between the TV and the laptop screen as they chat about tomorrow's weather and about Jeff's greatgranduncle.

However, the extension of evening family time could happen properly only on the weekends. During an interview, Marilyn mentioned some of the ways in which she and her family made family time feel longer and more special during the weekends. "At weekends, it's much more relaxing, so we might put on music, we might close the curtains a bit early, maybe have some candles. We might eat in there" (interview Marilyn).

In his work on ICT use with families in London in the early nineties, Hirsch (1998) finds that his research participants were often preoccupied with creating a sort of alternative time of unity or uninterrupted togetherness on the weekends. One family created this alternative time through a set of strategies that involved doing the shopping and the housework before the start of the weekend and a set of visible changes in everyday routines, such as, for the male adult, not shaving and removing his wristwatch on the weekend.

In a context of evening indulgence that feels cosy and that people would like to extend, the pursued activities generally followed individual interests and involved the use of individual digital media. The TV would be on, providing a common focal point and creating a soundscape and a "lightscape" that was shared, but the people around it constantly switched their attention from the TV content to their personal preoccupations. These preoccupations varied with the digital devices available to them. Dominic described in detail the "technological abundance" of their evenings, which was visually depicted in the collage made by him and his family:

Dominic: Last night, we had Ewan using the laptop with headphones on, sitting on one of the sofas. And then Justine and myself were playing games on the iPhones. Lara was watching the TV, and then she switched on to using the laptop. I suppose Justine and I were watching the TV at the same time, but we were playing on the iPhones as well, so yeah…

Roxana: Does this normally happen in the evenings then?

Dominic: Yeah, normally that's an arrangement. Ewan does normally play—if he can't be on the Xbox 'cause the TV is on, then he would quite often go to the laptop and look at YouTube and stuff like that with headphones on. And Lara and I would probably watch the TV, or one of us might be using the iPhone to do his stuff. And, when Justine is here, she's on the iPhone nonstop, she's continually using it: texting, Skyping, all sorts of things.

Roxana: She only chats on Skype?

Dominic: Sometimes she speaks with people as well, and it's quite annoying …so, the habit would tend to be that Justine would

| | probably be using either her iPhone, or her laptop, and the TV would be on 'cause she's the one doing it the most. She might put a film on TV, and have the laptop in front of her, and she would be Facebooking or Skyping while the film is running in the back. |

Roxana: And she's watching the film?

Dominic: She says she is, she says she is, yeah. And Ewan probably; he focuses a little bit more on one activity I think. So, the TV might be on, but if he's on the laptop with the headphones, he's generally looking at what he's doing on the laptop. But, if he sees something on TV, he might take his headphones off and then tune into the TV.

Ewan [from the sofa, while playing a video game]: If there's something interesting.

Dominic: Something interesting grabs his attention. Or, if something happens on his phone, he might pick his phone up. [to Ewan] You would usually phone less, don't you, then perhaps laptop? [Ewan doesn't reply] Lara would watch the TV, she might have her laptop open, but that's generally because she's doing some work as well. But she quite often does stuff on her iPhone while the TV's running. So, she's watching something, a TV programme, but she's also communicating with her mum, or a friend, or Justine, or something like that, on the iPhone. And I would generally watch the TV or read the newspaper; or if the TV wasn't very interesting and nobody's using their iPhone, I might borrow an iPhone and play a game on it. That's generally how it happens, you know. (interview Dominic)

The diversity of technology permitted flexibility and creativity in the activities pursued and in the media of the pursuance. Dominic, who did not have his own iPhone, was happy to spend his evening time switching between watching TV and reading the newspaper or a book, and he would sometimes borrow an iPhone (if available) to play a game. Lara might work on her laptop while watching TV, or she might use her iPhone to communicate with her mother who lives in Spain or with her daughter during term times when she is away. The possibilities are rich; people do not need to stick to the same activity or to the same device every evening.

This picture looks different from the situation described by Morley (1986) in his study of family TV watching in Britain in the mid-eighties. In Morley's study, the living room TV was the main source of media content, and the remote control device was generally used as an instrument of male dominance, which triggered conflicts over the choice of programme and over the viewing style. He outlines that the male research participants preferred attentive silent viewing, whereas the female participants liked to chat and to multitask while watching TV. This was not the case for the participants in my research, which was conducted more than twenty-five years later than Morley's. Dominic did not try to assert control over the environment of the living room by choosing what his family would watch. Chatting and partial attention to the TV screen were described as part of the ordinary watching style. This change was also supported by the multitude and the availability of media content: one could re-watch a programme anytime through the Catch Up function of TV boxes or on the internet, so there was no need to pay total attention at the moment of the first watching. The quantity of media content might also make it difficult to decide to watch a single programme if one's interests were diverse. Another difference from Morley's study is that the participants in my research placed an emphasis on spending time together in the same room rather than on having a family favourite programme that they would all watch attentively. As Vic put it, "Generally, we do all sit together and watch whatever's on, and chat at the same time, and play on iPads, and talk, and read. So, we do like to be in the same room." (interview Vic and Gail)

Even if the use of multiple new digital media might paint a picture of "technologically-saturated" homes, people sometimes described their media use in terms of continuities rather than changes. Vic said that using the iPad was "almost like browsing an electronic magazine," and his wife Gail agreed that there was nothing special in using the iPad while watching TV—it was just like "reading the newspaper while watching TV."

Returning to Dominic's description of their family evenings, there was a sense of flexibility and lack of normative expectations from the other family members. Everybody was left to pursue their own interests even if they might sometimes produce irritation, such as in the example of loud chatting with friends while the other family members were watching the TV. Ewan played games on his laptop, aurally isolated through his headphones. Justine texted and talked with her friends on Skype. Lara and

Dominic did "their stuff" on the iPhone or laptop. Diversity and individuality were encouraged. What I found surprising was Justine's choice to virtually bring her social circle into the living room instead of looking for privacy to chat with her friends. In addition, the acceptance by other family members of this activity—although Dominic found her loud talking annoying—made togetherness in the evenings possible. During university holidays when she was at home, Justine did not need to choose whether to actively spend her time with her family or spend it by chatting with her friends. She could do both activities simultaneously.

In a publication drawn from their research with children in the North of England about experiences of time at school and in family life, Christensen, James, and Jenks mention a preoccupation with keeping both family togetherness and individual independence at home. They outline that, in order to manage disputes between family members over time and space in the home, families carry out a process of "constant balancing of the concerns for independence, and that of togetherness, a process articulated in and around particular spaces in the home" (2000: 128).

From this perspective, evening times in the living room could be seen as daily opportunities for exercising this act of balancing togetherness and independence. I argue that family time in the evenings can thus be defined as a situation of physical togetherness and digital independence.

During the collage-making activity, the participants in my research often asserted individuality through their choice of the fabric that was used to express how time felt for them at different moments. For communal evening times, they often chose different fabrics, revealing that even if they were physically together and sharing the same room and time slot, their experiences of this shared event were different. For example, in the collage represented in Fig. 7.5, Jeff identified three distinct moments during his evening times: (1) 7 p.m. was illustrated by pink felt, (2) 10 p.m. was illustrated by white and cream patterned cotton, and (3) 11 p.m. was illustrated, again, by pink felt together with a cup of tea and snacks. This is how Jeff explained his choices and described these moments:

(1) Properly relaxed, switched off, just chilled out. I tend to have a bit of a dip, quiet time, and wake up again later. This one is the softest—the pink one is softer. (2) A little bit more awake but still a soft, nice material. Probably I would do a bit of work, something more active. On occasion, probably, I would do a bit of email, hobby related. (3) So, my final one is a sleepy one; might fall asleep on the settee and then wake up and go to bed.

(transcript from video of Tactile Time collage making with Cynthia and Rob's family)

In his description, Jeff's evening was experienced individually, even if it was part of a shared time and space. During the collage making, his daughter Emma reacted to his last entry (3) and to his choice of adding snacks and a cup of tea to illustrate it by saying, "What's that? Eat because we've gone away?" For Emma, Jeff's choice of snacks and tea expressed a celebration of the fact that he would spend some time alone "unsupervised" in the living room, which can be seen as a small "betrayal" to the ethics of togetherness in favour of individuality.

At her turn, Emma portrayed her evening by identifying three different moments: (1) 7 p.m. was illustrated by blue tulle netting fabric, (2) 8 p.m. was illustrated by white soft muslin together with a snack of toast and honey, (3.) 8:30 p.m. was illustrated by dark red felt, which other participants described as the least soft of all three types of felt provided. This was how Emma described these three moments and explained her choice of fabrics:

(1) I think I get quite excited around the nice programmes in the evening. (2) This felt really soft. I'm calming down a bit more, but I'm still watching the TV, and I still like it. So, I'm still awake. (3) Quite soft, like the bed duvet; still happy because I'm going to read my book. (transcript from video of Tactile Time collage making with Cynthia's family)

For Emma, the evening was about pursuing her interests: the evening TV programmes that she likes and her book that she reads in bed. Her choice of fabrics was related to the surfaces of the home—the red felt was soft like the bed duvet—or to other tactile textile experiences. Her mum Cynthia suggested during the collage-making that Emma's choice of tulle to express excitement could be related to what tutus meant for her: Emma attended a dancing club on the weekends, regularly taking part in shows that her group created and performed for an audience of family and friends (one of which I, too, attended together with Cynthia). Thus, through her choices of fabrics as well as through her evening activities, Emma expressed her individuality.

In the examples of Jeff and Emma, as well as in the previous example of Dominic, the family appeared as a context for individual pursuits. It can be a context that actually encouraged individual interests and experiences, or, at least, it did not conflict with them. I suggest that even if family time and the idea that the home is most homely when all family members are around were initially emphasized by all my participants, a closer look at what people did during family time shows a focus on individual pursuits. My ethnography shows that, in everyday life, "family" is not so much a normative social construction but is rather a context that lets or even encourages its members to find and express their individualities.

In a publication about changes and continuities in contemporary family, cited by Simpson (1998), Elliot (1986) identifies a "tragic paradox" in which "security in personal relationships implies commitment and loss of freedom" (1986: 133). The materials that I discussed in this chapter show that there are ways to dissolve this tragic paradox and that contemporary families are able to keep security and commitment, as well as a sense of freedom, at the same time. Some of the ways in which people attempt on an everyday basis, to dissolve this paradox is by using a variety of digital media at evening times.

Media Multi-tasking

In this chapter, in discussing the ways in which people employ digital technologies in order to create a specific temporal modality—family time—together with emergent forms of sociality, my focus was not on digital media *per se*. However, I believe that my ethnographic description can speak to other bodies of literature that look at family life in relation to (digital) media practices, such as media anthropology (Madianou and Miller 2011, 2013), cultural studies (Moran 2007), and media studies (Morley 1986; Moores 1993).

Moreover, this chapter can also be regarded as an ethnographic illustration and discussion of what has been recently identified as an important change in ICT domestic consumption. While the ownership of smartphones and tablets has been rising exponentially, with two third of UK adults now owning a smartphone and 54 % of households owning a tablet computer (Ofcom 2015), the ways in which these new devices are used in relation to traditional media, such as the TV, are often defined by multi-tasking. Almost every adult (99 %) recalled being engaged in two or more media activities at the same time, with watching live TV and making voice calls being the most popular multi-tasking combination. It has been suggested that the average adult in the UK spends over half of their waking hours engaged in media or communications activities. However, because some media activities are conducted simultaneously, a total volume of 11 hours 7 minutes of media and communications activities undertaken by an adult per day is squeezed into 8 hours 41 minutes (Ofcom 2014).

While the cited reports only identify the rising trend of media multitasking by drawing upon survey methods, the ethnographic description of, and theoretical framework for, situating the ways in which people multitask with media that were developed here could provide additional insight for scholars and policy makers interested in this phenomenon.

NOTES

1. For more information about the "conventional" structure of British meals, see Douglas and Nicod (1974).
2. My affirmation is not intended to dismiss the existent sets of criteria that "external" individual agents and institutions sometimes follow in "evaluating" families—together with local government interventions into the lives of families that they have identified as prone to "troublemaking"—but just to express the strength of the idea that "every family is unique," which I encountered very often during my fieldwork.
3. Occasionally, they compared their actual experience of family life to the experience they remembered from their childhood, and this was often to prove the superiority of the former above the latter.
4. It is interesting to note that these two options were considered to be mutually exclusive.
5. To illustrate aspects of family time that were only expressed in the Tactile Time collages in this chapter, such as the choice of fabrics representing specific moments of the day, I employ scan copies of some collages that have all been anonymized by using picture-editing software.
6. The exceptions are what are called bungalows—houses with only a ground floor—and houses with a loft conversion or three or more storeys. In the sample employed in this research, from a total number of eighteen families, only one had a loft conversion while all the others followed the classical model of two-storey houses.
7. See Attfield (1999) for a discussion about the ways in which the parlour has been incorporated into the open-plan living room.
8. Gullestad (1984) makes a similar observation about the matrimonial bedrooms in Norwegian block of flats suburbs in her work with working-class young women in a Norwegian town.
9. See DeVault (1991) and Valentine (1999).

REFERENCES

Andersen, Holly, and Rick Grush. 2009. "A Brief History of Time-Consciousness: Historical Precursors to James and Husserl." *Journal of the History of Philosophy* 47 (2): 277–307.

Attfield, J. 1999. "Bringing Modernity Home: Open Plan in the British Domestic Interior." In *At Home: An Anthropology of Domestic Space*, edited by Irene Cieraad. New York: Syracuse University Press.

Brembeck, Helene. 2012. "Cozy Friday: An Analysis of Family Togetherness and Ritual Overconsumption." In *Managing Overflow in Affluent Societies*, edited by Barbara Czarniawska and Orvar Löfgren, 125–140. Oxon: Routledge.

Carsten, Janet. 2000a. "Introduction: Cultures of Relatedness." In *Cultures of Relatedness: New Approaches to the Study of Kinship*, edited by Janet Carsten, 1–36. Cambridge: Cambridge University Press.

Carsten, Janet. 2000b. "Knowing Where You've Come From: Ruptures and Continuities of Time and Kinship in Narratives of Adoption Reunions." *Journal of the Royal Anthropological Institute* 6 (4): 637–653.

Carsten, Janet. 2004. *After Kinship*. Cambridge: Cambridge University Press.

Christensen, Pia Haudrup. 2002. "Why More 'Quality Time' Is Not on the Top of Children's Lists: The 'Qualities of Time' for Children." *Children and Society* 16: 77–88.

Christensen, Pia, Alison James, and Chris Jenks. 2000. "Home and Movement: Children Constructing 'Family Time.'" In *Children's Geographies: Playing, Living and Learning*, edited by Sarah L. Holloway and Gill Valentine, 120–134. London: Routledge.

Collier, Jane, Michelle Rosaldo, and Sylvia Junko Yanagisako. 1982. "Is There a Family? New Anthropological Views." In *Gender and Kinship: Essays Towards a Unified Analysis*, edited by Jane Collier, Michelle Rosaldo, and Sylvia Junko Yanagisako. Stanford: Stanford University Press.

DeVault, Marjorie L. 1991. *Feeding the Family: The Social Organization of Caring as Gendered Work*. Chicago: University of Chicago Press.

Douglas, Mary, and Michael Nicod. 1974. "Taking the Biscuit: The Structure of British Meals." *New Society*, 744–747, December 19.

Edwards, Jeanette. 2000. *Born and Bred: Idioms of Kinship and New Reproductive Technologies in England*. Oxford: Oxford University Press.

Edwards, Jeanette. 2004. "Incorporating Incest: Gamete, Body and Relation in Assisted Conception." *Journal of the Royal Anthropological Institute* 10 (4): 755–774.

Edwards, Jeanette. 2008. "'Creativity' in English Baptist Understandings of Assisted and Assisting Conception." In *Creativity and Cultural Improvisation*, edited by E Hallam and Tim Ingold, 167–185. Oxford: Berg.

Edwards, Jeanette, S. Franklin, E. Hirsch, F. Price, and Marilyn Strathern. 1993. *Technologies of Procreation: Kinship in the Age of Assisted Conception.* Manchester: Manchester University Press.

Edwards, Jeanette, and Marilyn Strathern. 2000. "Including Our Own." In *Cultures of Relatedness: New Approaches to the Study of Kinship*, edited by Janet Carsten, 149–166. Cambridge: Cambridge University Press.

Elliot, Faith Robertson. 1986. *The Family: Change or Continuity?* Houndmills: Macmillan.

Finch, J. 1997. "The State and the Family." In *Families and the State*, edited by S. Cunningham-Burley and L. Jamieson, 29–44. London: Palgrave.

Gabb, Jaqui. 2010. *Researching Intimacy in Families.* New York: Palgrave Macmillan.

Giddens, Anthony. 1981. *A Contemporary Critique of Historical Materialism – Vol. 1, Power, Property and the State.* London: Macmillan.

Gillis, John. 1996. "Making Time for Family: The Invention of Family Time(S) and the Reinvention of Family History." *Journal of Family History* 21 (1): 4–21.

Gullestad, Marianne. 1984. *Kitchen-Table Society.* Oslo: Universitetsforlaget.

Hirsch, Eric. 1998. "Domestic Appropriations: Multiple Contexts and Relational Limits in the Home-Making of Greater Londoners." In *Migrants of Identity: Perceptions of "Home" in a World of Movement*, edited by Nigel Rapport and Andrew Dawson. Oxford: Berg.

James, William. 1890. *The Principles of Psychology*, vol. 2. New York: H Holt & Co.

Kremer-Sadlik, T., and Amy L. Paugh. 2007. "Everyday Moments: Finding 'Quality Time' in American Working Families." *Time and Society* 16 (2–3): 287–308.

Long, Nicholas, and Henrietta L. Moore. 2013. "Introduction: Sociality's New Directions." In *Sociality: New Directions.* Oxford: Berghahn Books.

Madianou, Mirca, and Daniel Miller. 2011. *Migration and New Media: Transnational Families and Polymedia.* London: Routledge.

Madianou, Mirca, and Daniel Miller. 2013. "Polymedia: Towards a New Theory of Digital Media in Interpersonal Communication." *International Journal of Cultural Studies* 16 (2): 169–187.

Miller, Daniel. 1998. *A Theory of Shopping.* Cambridge: Polity Press.

Moores, Shaun. 1993. *Interpreting Audiences: The Ethnography of Media Consumption.* London: Sage.

Moran, Joe. 2007. *Queuing for Beginners: The Story of Daily Life from Breakfast to Bedtime.* London: Profile Books.

Morgan, David H.J. 1996. *Family Connections: An Introduction to Family Studies.* Cambridge: Polity Press.

Morley, David. 1986. *Family Television: Cultural Power and Domestic Leisure.* London: Comedia.

Ofcom. 2014. *The Communications Market Report.*
Ofcom. 2015. *The Communications Market Report.*
Schneider, David. 1968. *American Kinship: A Cultural Account.* Englewood Cliffs: Prentice-Hall.
Schneider, David. 1984. *A Critique of the Study of Kinship.* Ann Arbor: University of Michigan Press.
Silva, Elizabeth Bortolaia, and Carol Smart. 1999. "The 'New' Practices and Politics of Family Life." In *The New Family?*, edited by Elizabeth Bortolaia Silva and Carol Smart. London: Sage.
Simpson, Bob. 1997. "On Gifts, Payments and Disputes: Divorce and Changing Family Structures in Contemporary Britain." *Journal of the Royal Anthropological Institute* 3 (1): 43–59.
Simpson, Bob. 1998. *Changing Families: An Ethnographic Approach to Divorce and Separation.* Oxford: Berg.
Stacey, Judith. 1990. *Brave New Families: Stories of Domestic Upheaval in Late-Twentieth-Century America.* Berkeley, CA: University of California Press.
Strathern, Marilyn. 1981. *Kinship at the Core: An Anthropology of Elmdon, a Village in the N-W Essex in the 1960s.* Cambridge: Cambridge University Press.
Strathern, Marilyn. 1992. *After Nature: English Kinship in the Late Twentieth Century.* Cambridge: Cambridge University Press.
Valentine, Gill. 1999. "Eating in : Home, Consumption and Identity." *The Sociological Review* 47 (3): 491–524.
Williams, Fiona. 2004. *Rethinking Families. Moral Tales of Parenting and Step-Parenting.* London: Calouste Gulbenkian Foundation.

CHAPTER 8

Saving Energy in British Homes: Thoughts and Applications

I will now go back to the context of applied interdisciplinary work that framed my research to show how my ethnographic findings, which were influenced by this context, could be applied to change the type of questions that projects focused on energy reduction address.

I will start with situating this work in relation to approaches to interdisciplinarity and applied anthropology. Following Strathern's (2006) discussion of interdisciplinary research, I will argue that the potential of interdisciplinarity emerges *a posteriori* the completion of disciplinary research when one might let oneself "be captured" by concerns and formulations from other fields while and *through* maintaining a critical stance.

The ways in which research questions were formulated in previous applied studies of domestic energy consumption will then be discussed. Following the work of the geographer Kersty Hobson (2011), I will argue for a change of perspective from a "homes as places where people consume energy" premise to a "homes as sites of action" approach.

Subsequently, I will introduce several ideas for potential applications of the ethnographic findings that were the focus of the empirical chapters. First, I will look at the ways in which the idea of opting for a family-style lifestyle in English middle-class kinship (Strathern 1992) organizes formulations of morality, and I will ask whether and how everyday actions oriented towards the preservation of the environment can fit into this system of thought. Second, I will discuss the existent conflicts between the

© The Editor(s) (if applicable) and The Author(s) 2016   171
R. Moroşanu, *An Ethnography of Household Energy Demand
in the UK*, DOI 10.1057/978-1-137-59341-2_8

folk models of time and agency employed in everyday domestic life and the temporality expressed by the Climate Change Act. Potential ways to reconcile these distinct approaches to, and enactments of, time will be proposed as a second set of applications.

INTERDISCIPLINARY RESEARCH AND APPLIED ANTHROPOLOGY

To situate my doctoral research in relation to the interdisciplinary applied project that framed it, and to a certain extent made possible, I will focus on two elements of this context that were essential in influencing the research design, the completion of my ethnographic fieldwork, and the analysis of my findings: interdisciplinary discussions and the expectation of applications.

Interdisciplinarity, in its contemporary understanding as a way of conducting research and producing knowledge that crosses the boundaries between physical sciences and humanities and social sciences (Barry et al. 2008), has recently "come to be seen as a solution to a series of contemporary problems, in particular the relations between science and society, the development of accountability and the need to foster innovation in the knowledge economy" (idem, p. 21). When it does not happen naturally, the creation of interdisciplinary endeavours is encouraged by funding bodies; in funding calls, and in subsequent findings reports, the evocation of the degree of interdisciplinarity that has been achieved could often be regarded as representing an index of accountability and of innovation in itself (Strathern 2004).

Marilyn Strathern (2004, 2006) questions the assumptions that this new form of knowledge production is based upon. She distinguishes between a managerial model of knowledge creation and a research model, and she discusses their characteristics in relation to the contemporary context where "evidence-based policymaking goes hand in hand with a Euro-American understanding of the world as full of uncertainties … [that] are not just political or economic but epistemological: we do not know enough—more research is needed" (Strathern 2006: 193). While a research model is concerned with understanding actions and processes for their own sake, research in a managerial model is employed to seek normative propositions to be followed in giving advice on further action. Strathern outlines that she does not oppose the two models in totality, and she admits that they are often employed in combination in today's academic context. She suggests that each of the two models should be seen as representing a specific point in "a particular Euro-American oscillation between the condition of knowing through investigation (research) and the condition of asking what is to be done with that knowledge (management)" (2006: 195). From this perspective, the project that framed my doctoral research can be seen as expressing a managerial model of knowledge creation where knowledge is produced with the intent of informing future technological and policy-based interventions in relation to a reduction in domestic energy demand. These potential future applications of the knowledge that the project was expected to deliver are part of the wider management process of reducing the UK's carbon emissions by the year 2050, which is legislated by the Climate Change Act.

When the research carried out by following a managerial model is also defined as interdisciplinary, a main problem that can arise, as outlined by Strathern, is the criteria for evaluating it. Interdisciplinarity is something more than a discipline; it is a dialogue between disciplines. More than a set of methods and a field of problematics, for Strathern a discipline is also "a bundle of yardsticks, that is, criteria for evaluating products and maintaining standards" (2006: 199). An important framework for evaluation offered by a discipline is criticism. In contrast, interdisciplinary endeavours cannot be evaluated inside similar frameworks: "The single outcome, the integrated collaboration, is impossible to measure against its own diverse origins" (2006: 201). In this situation, Strathern sees the scope of interdisciplinarity to be in the possibility of initiating criticism as a way of letting oneself be captured by the concerns of a different discipline. Therefore, the potential of interdisciplinary work appears after the completion of the disciplinary research, at the point when it is possible to shift from a research model of knowledge production to a managerial model to ask what is to be done with the emergent knowledge.

The second characteristic of the interdisciplinary context that was instrumental in the development of my research was the applied dimension. The position of working as part of an applied project as an anthropologist brings up discussions about taken-for-granted distinctions between "applied anthropology" and "academic anthropology"; this was discussed in more detail in the methodological chapter. Sillitoe (2007) argues that these two forms of anthropological research should not be regarded in opposition and suggests "academic research is often a prerequisite for flourishing applied work" (2007: 161). In order to move the discussion forward from a dichotomous conflict to a form of reconciliation that would contribute to the advancement of both types of endeavours, he asks how the discipline of anthropology can be applied. He suggests that there are two ways in which the idea of applying anthropology can be understood: applying the methods of anthropology and investigating the applicability of the knowledge that anthropologists have learned through their fieldwork and have "systematized in various ways using the theories of the moment" (2007: 155). Both approaches could be equally fruitful even if they are based upon different sets of findings. Applying anthropological methods to the study of contemporary questions can bring a fresh perspective to a set of research interests that have previously been conscientiously delimited and defined, such as the concern with the ways in which people interact with energy metres (Hargreaves et al. 2010). Nevertheless, this approach can often be expected to be articulated in a problem-solving manner in the context of interdisciplinary energy research projects in which anthropologists are often regarded as "people experts" (Henning 2005).

Applying the knowledge that anthropologists have gained through long-term ethnographic fieldwork is a process that necessitates a wider timeframe (Rabinow et al. 2008). First, it requires one to have the opportunities to engage in long-term fieldwork and to pursue an open and flexible research agenda that would make it possible to follow serendipitous findings and relationships. Second, it requires one to have the readiness, after analysing this knowledge in relation to a corpus of anthropological theory, to look back at it again from the standpoint of an applied agenda. In this more "traditional" approach to ethnographic fieldwork, one is required to momentarily suspend any concerns with applications and to engage in a learning process that would unfold the relevant ideas, meanings, and processes in relation to which research participants organize and make sense of their lives. This knowledge corresponds to a research model, in Strathern's (2006) conceptualization, and is legitimate in itself. The reinterpretation of this knowledge from the standpoint of an applied agenda would be a whole different process and would produce a new corpus of knowledge. In my work, it is this form of applied anthropology that I seek. After having interpreted my findings in the previous chapters in relation to existent anthropological theory, this last chapter represents the beginning of a process of secondary analysis that follows concerns with possible applications in relation to the domestic sustainability agenda.

QUESTIONING THE POSITIONING OF DOMESTIC ENERGY CONSUMPTION RESEARCH

In the introduction to their collective volume on the anthropology of energy and cultural ideas about energy across the contemporary world, Strauss et al. (2013) outline that "because of the necessity of institutions to manage energy flows, and because of the necessity of energy flows to individual agency, an anthropology of energy is necessarily political" (2013: 12). Institutions that regulate energy markets and create systems of transaction for energy resources and carbon emissions are, arguably, the main actors in changing "energyscapes" (2013: 11) at national and global levels. However, in public-funded research projects focused on reducing energy demand, these institutions are sometimes taken for granted. They represent the context that made the research possible, and the results of the research need to be "delivered" to them in order to inform further forms of planning and regulation.

As a consequence, researchers working in what could be defined as one-sided projects oriented only towards the practices, beliefs, and behaviour of laypeople can be regarded by research participants as representatives and "translators" of the national environmental agenda to local communities and domestic consumers in a top-down perspective. As Hobson (2003, 2011) identifies, which is also the case in my research (see the example from Chap. 4 with Chris's reaction during a project workshop), the actual agenda of domestic sustainable consumption is often considered unjust by laypeople. They express the feeling that, by changing their domestic practices and by "doing their bit," they are not making any real difference if the forms of action that the government takes are inefficient.[1] Thus, if publicly funded researchers can sometimes be oblivious of the political entanglements that frame the domestic sustainability agenda, everyday people are not. I argue that when working as part of a project that promotes domestic sustainability, critical awareness of the context that frames the research needs to be part of the process of interpreting one's findings. This does not necessarily require a change in perspective from the actions of laypeople to the processes through which the environment was formulated as a political project (Macnaghten 2003; Urry 2011). It does require, however, questioning one's assumptions in designing and carrying out a research project that responds to the domestic sustainability agenda. I follow this argument by focusing on two strands in the formulation of the domestic sustainability agenda in research and policy. In this section, I look at how domestic space has been conceptualized in literature that reports on findings of applied projects that investigate household energy demand. In the following sections I will address the temporal feature of the environmental policy agenda in the UK, as expressed in the Climate Change Act, in relation to the temporalities employed by research participants in their everyday lives.

The field of domestic energy research can be divided into two main approaches in relation to the epistemological and methodological assumptions that are articulated when scholars define domestic settings as either "houses" or "homes" (Ellsworth-Krebs et al. 2015). While scholarship focused on "houses" follows physical sciences paradigms and methods for looking at the physical characteristics of buildings where the occupants are often assumed to be passive, a new research agenda focused on "homes," Ellsworth-Krebs et al. argue, would tackle the real challenge of developing new understandings of energy demand by combining physical and social factors. Nonetheless, it can be argued that—and as the previous chapter illustrated—the meanings attributed to the notion of household in political discourse and people's everyday experiences of domesticity are two completely different things. The geographer Guy Hawkins observes that "the household invoked in environmental policy is highly normalized and constituted through specific empirical processes. In contrast to this, 'home' emerges as a complex spatial and temporal field where everyday life unfolds" (Hawkins 2011: 69). In other words, the "domestic" in the "domestic energy demand reduction" agenda and what actual people refer to as "home," are not the same entity. The former represents the level of residential consumption, which in addition to the energy consumption of the commercial, public, and industrial sectors, gives the equation of the national levels of energy consumption and carbon emissions. The latter is a site of dwelling where people carry on their lives and attribute meaning to their experiences, sometimes in the company of others. In order to link the abstract notion of the household, which has "appropriate" levels of energy consumption attributed to it, to the idea (Douglas 1991) and the experience of home, an approach to homes as "places where people consume energy" has been often adopted in interdisciplinary research on energy consumption as a premise to build further investigations upon.

While providing a good-enough framework for linking physical and social sciences approaches to obtain research funding, this premise could considerably limit the interpretation of one's findings to an area of application that assumes the household as a site for government intervention and regulation. This perspective expresses a vision of social change as a unidirectional process that is introduced by "structural players" (Middlemiss 2010), such as governments, in order to subsequently be adopted by individuals by changing their habits and behaviour in the face of societal "external" transformations. In this vision, the cultural domains of private-domestic and public-"politico-jural" (Fraser 1990; Yanagisako 1979; Yanagisako and Delaney 1995) appear as essentially divided in relation to potential enactments and expressions of political power.

As Hobson (2011) argued, scholars adopting Foucault's ideas on governmentality and biopolitics have criticized the discourses and intentions of governmental interventions related to the domestic sustainability agenda for trying to impose a vision of "a self-reflexive individual taking responsibility for knowing and reducing his or her emissions" (Rutland and Aylett 2008: 642) and for aiming to create a "responsible, carbon-calculating individual" (Slocum 2004: 765). The research focus on the reduction of domestic energy demand was also challenged, Hobson (2011) shows, by scholars who chose instead to look at community energy projects that, they argued, can be seen as sites of agency, in other words, "a means for political activity on the part of the broad mass of citizens who join not just for social interaction but also to be actively involved in the making of public policy" (Hoffman and High-Pippert 2005: 399).

Hobson suggests that there are ways of looking at domestic life that do not abandon the possibility of a form of personal environmental politics in favour of green governance techniques. She proposes an approach based on the work of Foucault (1990, 2000) on ethical practices, suggesting that "one could develop one's own ethics as a way of resisting governmental imperatives" (Hobson 2011: 204) through forms of self-reflection and by constantly asking "What do I aspire to be?" (Cordner 2008). Her argument, which was discussed in more detail in Chap. 2, represents an essential point of reference for the way in which I orient my work in relation to the domestic sustainability research agenda. In the perspective advocated by Hobson, the domestic becomes a site of political power; it is no longer the counterpart of the public-"politico-jural" domain (Fraser 1990) but its continuation. Following this argument, one can replace the "homes as places where people consume energy" premise with a "homes as sites of action" approach. It is the aim of this work to follow the latter approach.

One way in which this perspective can be employed in interdisciplinary research on domestic energy consumption is by provisionally suspending a direct concern with energy demand to look at the forms of actions people do inside their homes that make them feel empowered—

actions that are, in Foucault's (1990, 2000) terms, "practices of freedom." During my fieldwork, I discovered it is these actions that people value the most; this insight is translated here into the concept of "ordinary agency." Strauss et al. argue that "how people use energy is related to how people value it; and how people value energy is related to what it enables them to accomplish not only materially but also socially and culturally" (2013: 15). By looking at the forms of ordinary agency that are articulated and enacted inside the home, one momentarily loses the focus on domestic energy demand only to recover it later, enriched and able to respond to questions that are not merely following a problem-solving approach.

By adopting the premise of "homes as sites of action" and, most importantly, by proposing it at a wider level in policy formulations, a "constructive and positive slippage" (Hobson 2011: 205) can begin to happen with individuals and societal institutions overcoming a crisis of trust (Giddens 1994; Beck and Beck-Gernstein 2002) to engage together towards enacting a more sustainable future. Some ways in which this collaborative approach can be expressed are by mediating between the temporalities of institutions and the temporalities of everyday domestic life, as I will discuss later.

FAMILY-STYLE LIVING AND ENERGY CONSUMPTION

When the Low Effort Energy Demand Reduction (LEEDR) project advertised for participants, they called for "home-owner families." In this description, neither a suggestion of class nor of demographic characteristics—such as the size and "type" of family (nuclear, extended, or single-parent), the civil status of the couple, or their sexual orientation—were present. Still, the families who volunteered, those who defined themselves as "home-owner families" and who were interested in taking part in the project, were all families in which the adults were heterosexual and married and who had children. With the exception of a single-parent family and of an extended family that included a live-in maternal grandmother, all of the families were two-parent nuclear families. From the twenty family participants, eighteen could be described in terms of income and education as "middle-class," whereas the other two could be described (and self-defined during interviews) as "working-class." While this might largely express a characteristic of the provincial locality where the research was set, it also shows that regardless of how diverse contemporary forms of cohabiting might be, when it comes to represent one's cohabiting group as a family to a team of public-funded researchers—who, by addressing the questions formulated by a public funding body, might be seen as representing the national government—a "traditional" representation of family takes precedence before other alternative representations.

What is it that makes the construct of family in the UK appear and be expressed as "enduring," as a "pillar of supposed stability" (Silva and Smart 1999: 2–3) in the face of various other social changes? As discussed in greater detail in the previous chapter, I follow the theoretical apparatus developed by Strathern (1992) to examine English middle-class kinship in my work. According to her, the "first fact" of English kinship is "the individuality of persons" (1992: 14). The articulation of the concept of family can thus be regarded as following what Macfarlane (1978) calls English individualism and describes as being an important cultural trait dating back centuries before the beginnings of industrialization. English individualistic ideas suggest that, through domesticity, one is legitimated to lead a form of self-supporting and self-sufficient lifestyle that limits the amount of close relationships with non-kin. Strathern (1992) suggests that later, in what she calls "the modern epoch," with the appearance and development of the middle classes, the domestic space came to stand for personal cultivation and individual improvement. "The internal (what is within persons) has been literalised as an interior (residential) space" (1992: 103). Furthermore, in post-Thatcherite Britain, individualism, as both self-improvement and self-sufficiency and privacy, is achieved through family-style living as an experience and a display of right choosing and acting. The ideas that "the quality of home life has an independent measure" (1992: 149) and opting for a family-style lifestyle shows one's capacity for morality, together with the assumption that people naturally know what is right by having had internalized convention as personal style, reveal families as distinct universes following internal self-designed and self-imposed norms.

During my fieldwork, the people who took part in my research used the concept of family in both descriptive and explanatory ways. They used it to describe the form of social organization that organized their experience of domesticity—a family-style lifestyle, in Strathern's words. They also used it to explain the reasons why they did specific things in their house, such as leaving on the landing light overnight, or the reasons why they did things in specific ways. For example, several families explained that they would normally use the tumble dryer to dry their laundry *because* they are a family (of four or more) and they have a lot of laundry. Thus, family was understood to be a self-sufficient entity that had both agency of its own—family as a domestic "reality" that asked for specific goods and practices such as a tumble dryer, in order to continue to exist—and that collected the agency of individual family members in a concentrated form, with people engaging in "family time" routines or creating an "elusive presences" form of sociality so that they would "do" (Morgan 1996) family. However, if a lifestyle can have agency in itself, then who or what is responsible for (the consequences of) the practices and commodities this lifestyle entails? It is neither the lifestyle nor the individual persons who choose it and who, by choosing it, have already shown their capacity for morality, as I discussed in the previous chapter. This is, therefore, a complex societal problem that needs to be addressed in much broader ways than just by asking families to reduce their energy consumption *and* still stay families.

In his study of British green communes from the 1970s, the geographer David Pepper shows that communal living appeared and developed as an alternative to the nuclear family, which was considered a form of social organization that was "unnaturally exclusive" (1991: 10) and represented a terrain for gender inequalities and exploitation. As an individualistic unit oriented towards itself, the nuclear family was seen as incapable of generating and promoting an altruistic way of living that would account for the needs of the environment and would let all its members feel equally empowered.

Concerning this issue, there still seems to be an irreconcilable everyday dilemma between the ideas of caring for one's family and caring for the environment. During the first LEEDR interview with Cynthia and Jeff's family conducted by a researcher from the discipline of design that I accompanied, Jeff mentioned that his brother used to live off the land in a commune until his first baby was born. After this event, he and his partner moved back to a town where he got a job and constant income, which allowed him to look after the baby's material needs much better than he could have done by only counting upon cultivating the land as part of a commune. The conclusion of this story, which linked this narrative with other similar stories that I came across during my fieldwork, was the unquestionable fact that, once one has a baby, going back to live as part of the established social order was the best choice. Cynthia expressed this dilemma in relation to their use of the tumble dryer. She said that they would like to be able to use the machine much less frequently than they do at the moment, but when school uniforms, work clothes, or other frequently used items needed to be washed and dried by the next morning, they did not have a choice but to use the tumble dryer. Self-sufficiency in doing the laundry, as opposed to the situation when one needs to rely upon the weather to get the laundry dried, was linked to normative expectations related to living as part of the established social order and opting for convention as a way of expressing right acting and one's capacity for morality. When faced with an ambiguous action, which did not represent the "greenest" or the "lowest-carbon" choice, people emphasized the positive role this action had upon their family life, where keeping up the routine was seen as positive in comparison to a potential disruption that a demand to account for the needs of other entities besides their kin could have brought.

In her monograph of a small Norwegian town, Norgaard looks at what she calls the socially organized denial of climate change in everyday life, which she defines as "the process by which individuals collectively distance themselves from information because of norms of emotion, conversation, and attention and by which they use an existing cultural repertoire of strategies in the process" (2011: 9). An important such cultural strategy that she identifies is based on local cultural homogeneity and on a strong emphasis placed upon tradition, which provides a normative definition for the concept of being a "good person" as well as a set of "proper" ways to do any type of everyday activity. The way denial is organized, Norgaard (2011) argues, can differ with the cultural context.

In my research, it was not a process of denial as such that I discovered to be instrumental for people's responses to the idea of climate change. The participants in my research were sensible to environmental concerns, and many of them engaged in a variety of everyday practices of sustainability such as growing their own food, cycling to work, or using green energy produced by the solar panels that they had installed on their roof. But, when everyday choices had to be made between an action with good (long-term) consequences for the environment and an action whose (immediate) consequences would suit one's family, the family was considered more important than the environment—or in Strathern's (1992) conceptualization, "culture" was considered more important than or superior to "nature". In other words, in the provincial context where I carried out my fieldwork, the normativity carried by the concept of middle-class family appeared to be much stronger and much more widely adopted than possible norms of caring for the environment.

Concerning the second problem of the nuclear family identified by Pepper's (1991) research participants, that of being a form of social organization based on inequality and on everyday forms of domination and oppression, my fieldwork showed a different story. As discussed in the previous chapter, the people who took part in my research were concerned with creating forms of sociality that involved both togetherness and independence, thus contributing to "doing" family (Morgan 1996) and encouraging individual interests and preoccupations at the same time. These actions expressed people's "ethical imagination" (Moore 2011) in developing relationships of mutual empowerment. The results of Pepper's study show that life in a commune does not eliminate relationship problems and the unequal distribution of power. He suggests that, by living in communes, people "merely swap one set of family problems (nuclear) for another (extended)" (1991: 151). The potential change of focus that might have happened in the last two decades between Pepper's research and mine, from a perspective on commonality and individualism as mutually exclusive to a concern with achieving and experiencing togetherness and independence simultaneously, shows that the nuclear family is not a form of social organization that is oppressive in itself. Rather, people are able to make this form of co-habitation into what they wish it to be.

The problem, therefore, is not that the nuclear family is the dominant form of social organization in the English context. The problem is that the middle-class family is equated to a dominant form of morality that is contained in its very articulation: the morality of having a family-style lifestyle and bringing up children that is sufficient in itself and does not leave room or time for other ethical concerns oriented towards entities other than kin. By drawing upon Strathern's (1992) work on English middle-class kinship, I suggested in the previous chapter that if having a family-style lifestyle embeds convention and if convention is seen to express morality, then opting for a family-style lifestyle shows one's capacity for morality and for right acting. The fact that concerns with the environment and with one's kin were often articulated in mutual opposition should not be seen as expressing a fundamental conflict, but only as illustrating one way in which these ideas were connected in the domain of everyday life at one particular point in time. In making and explaining various everyday choices, such as those discussed earlier, the participants in my research made appeals to the morality embedded in having a family-style lifestyle because it represented the principal moral system they could communicate to a team of researchers interested in reducing domestic energy demand in a secular, middle-class, provincial context. Making a set of everyday choices that would favour the environment over one's kin's immediate comfort would be associated, instead, with alternative lifestyles that promote communal living and a return to the land. But, as these alternative lifestyles are *alternative* and do not embed convention, they do not have the capacity to express morality in the same way as the idea of family-style living does.

This suggests that it would be difficult to reach a growing concern with the environment if a set of normative ideas about family-style living as inherently moral is maintained and promoted at the same time. Thus, when communicating research findings, it is not just that wider audiences would need to be given more information about the environmental impact of their everyday choices, but that the assumption that a family-style lifestyle embeds morality by default, in superiority to, any other possible lifestyle should be publicly and openly questioned.

Domestic Time and the Climate Change Act

I will now discuss the possible applications that the work developed in Chaps. 5 and 6 of this monograph, on the temporal modalities of spontaneity and anticipation, could have.

I will first go back to the way in which Greenhouse (1996) conceptualized a relationship between cultural models of time and understandings of agency. By carrying out a detailed analysis of previous anthropological approaches to the topic of time, Greenhouse suggests that there are mainly two forms of time, both geometric, that are employed in ethnographic writing: the cyclical, which is identified in ideas about nature, rituals, and the sacred; and the linear, which is used to understand progress and sequence (Fig. 8.1). She argues that the universalization of linear time as an objectively true construction of time served to naturalize particular forms of social order. In assuming linear time to be more "real," there is an inherent legitimation of the dominant contemporary institutional forms. However, one should not assume that this temporal shape organizes all forms of human experience, (such as imagination, taking action on impulse, or forms of domestic sociality) in all cultures. "Linear time, which is the cultural preserve of national histories and the public institutions comprised in them, dominates those institutional settings only to the extent that the people who inhabit them (whether managers or employees, presidents or citizens) believe that linear time has a transcendent reality

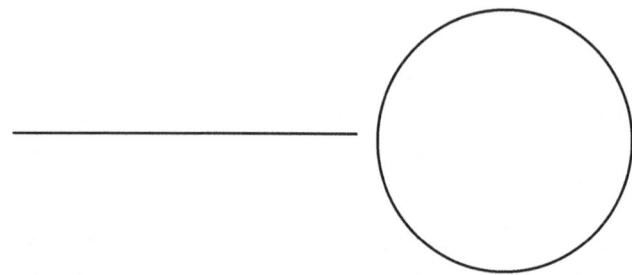

Fig. 8.1   Linear time and cyclical time (my drawing)

that allows it to absorb all the other, potentially rival temporal idioms that also suffuse daily life" (Greenhouse 1996: 209). An important role of the ethnographic endeavour would be, in such contexts, to describe and conceptualize some of the multiple formulations of time and agency that are not necessarily recognized as such by dominant institutional forms.

Here, I have identified and described spontaneity and anticipation as cultural models of time and as forms of agency that are enacted as part of the routines and contingencies of everyday life. I have argued that these are forms of "ordinary agency," related to "non-conscious" (unconscious, tacit, and non-intentional) acts of creating meaning, such as imagination and taking action on impulse. It was suggested that the value of these forms of ordinary agency does not stand in producing conspicuous and immediate effects on society. The ordinary agency engendered by spontaneity consists in generating feelings of empowerment over the people who enact it; these could subsequently be conditions for social change as they extend one's ethical imagination (Moore 2011) regarding one's relationships with oneself and with others. Similarly, anticipation can be seen as triggering the agency of the possible. The imagination of one's future agentic engagement with the world is, in itself, a form of agency. This ordinary agency, also expressed in actions of hope (Miyazaki 2004), refers to the "meanings people attach to questions of possibility" (Greenhouse 1996: 183) rather than to individual self-directed actions.

As temporal modalities, spontaneity and anticipation show different ways in which one could play with, blend, and knead the present and the future together. Spontaneity is an event where a present want and a future fulfilment are uniquely overlapped in the performance of a "happy action" (Taylor 1979). Anticipation is the act of making the future present by imagining a possibility, and it can be employed endlessly during a period of waiting for a future event. Here, the direction of time can be reversed so that, in moments of anticipation, the future travels back to the present; then it travels forward to its former place; then it travels back to the present once more whenever anticipation is employed again. Below, by following the conventional geometrical model expressed in the conceptualization of linear and cyclical time (Fig. 8.1), I propose a visual representation for spontaneity (Fig. 8.2) and for anticipation (Fig. 8.3).

The way in which I drew spontaneity represents the basic movement that is used in sewing and in waving; this also looks like the first half of the action of making a knot. Spontaneity happens as fast as all these three types of actions, and its results can be seen immediately similarly to the way in which the result of only one stich is immediately visible in the layout of an embroidery canvas. Spontaneity is, therefore, an effortless and organic way of modifying the reality, of changing the world.

Anticipation is a back and forth movement between present and future, which can also involve the past, for example, in anticipating the next Christmas one could remember other past Christmases that they experienced. In contrast to spontaneity, which depicts a moment, the back and forth movement that is anticipation can be the underlying temporal pattern of a whole day or of longer periods of time, for example, the year or

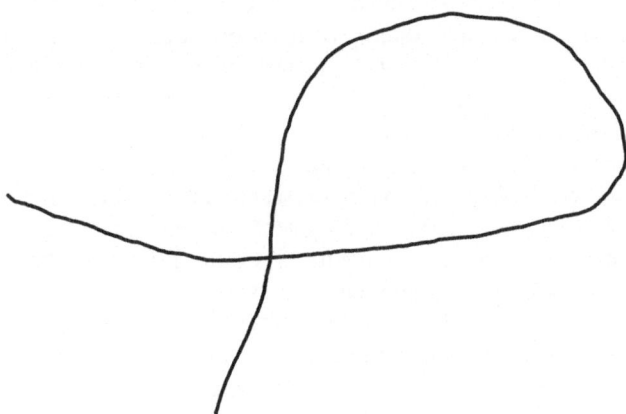

Fig. 8.2 Spontaneity (my drawing)

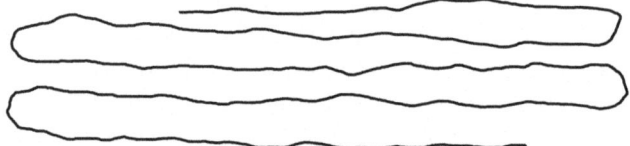

Fig. 8.3   Anticipation (my drawing)

more required to write a doctoral thesis when one anticipates the end of this process to take the shape of a finite material object. Anticipation is a means of living not only in the present, and it expresses the complexity of the domain of everyday life by emphasizing "temporal diversity" (Albert 2002; Geissler 2002), the multiplicity of "times" (Adam 1995), or "multitemporality" (Serres and Latour 1995).

Both spontaneity and anticipation can be employed as temporal modes in themselves or in combination with other temporalities, such as linear or cyclical, which are used and expressed by other individual agents and institutions. Separately or in diverse combinations, they are all part of the multiplicity of temporalities that exist concomitantly in one's life. Some of these, such as linear time, can be seen as the dominant temporality of the social structure in Western societies (Greenhouse 1996). Having been institutionalized as the time of public life, linear time is also the model invoked in some capitalist societies in relation to planning, ideas of efficiency, and specific systems of mass production, such as the "on-time" model (Hochschild 1997). A "well-planned time-saver schedule" attitude towards time related to the world of work in the USA was identified by Hochschild to be spreading inside the domain of the home in the way her research participants tried to efficiently organize their family time. However, she suggests that her research participants wished for "not simply more time, but a less alienating sense of time" (1997: 52), in other words, they wanted to approach time at home differently than at work. As discussed in Chap. 5, spontaneity expresses a different attitude towards time. As the ability to do something unexpected, spontaneity liberates people from the constraints of efficiency related to a neoliberal context and actions of "making" by expressing instead ordinary agency in forms of doing. When enacted within a linear temporal framework—the time of public life that organizes experiences of work and general activities of scheduling in relation to wider social systems and institutions—spontaneous actions can be regarded as directed towards interrupting the uniform temporal linearity in order to create wished for future moments and make them happen. Applying spontaneity to linear time would generate a new geometrical shape, represented below in Fig. 8.4. I propose to regard this drawing as an advancement of existent understandings and representations of linear time, which will show that people can act upon linear time in order to make it complex and eventful.

Similarly, when anticipation is used in relation to a cyclical temporal framework, such as when one anticipates the Monday pub quiz or the French dancing that takes place every second Wednesday of the month, the form of temporality generated could take the shape of a spiral (Fig. 8.5). The drawings above should be seen as a creative exercise meant to open anthropological imaginations towards other forms of temporality beyond the linear and the cyclical rather than as a "scientific" representation.

While I describe spontaneity and anticipation as cultural models of time that emerged from my fieldwork with English families in domestic settings located in a small provincial town, my research—as well as my identity as a researcher that the relationships with my participants have been built upon—can be regarded as situated within a different temporal framework set by national environmental policy. Linear time is the temporal form that underlies the environmental concerns in the UK that the LEEDR project, which frames the context of my research, responds to. These concerns are expressed in the Climate Change Act, a document set by the UK's Parliament in 2008. This document legally binds the Parliament to reduce the country's carbon emissions by 80% by the year 2050 in relation to the 1990 baseline. By setting this target, the Climate Change Act creates a linear timeframe where the solution to a present problem is left to a distant future moment of hopeful revelation. Both of the numbers used in the act, a reduction of 80% and the year 2050, are high/far in relation to the present; they seem

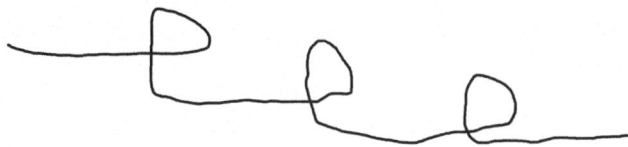

Fig. 8.4 Spontaneity and linear time (my drawing)

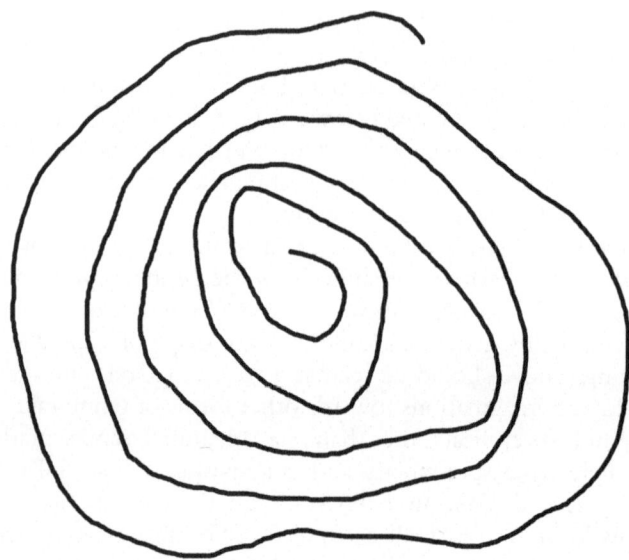

Fig. 8.5   Anticipation and cyclical time (my drawing)

unrealistic and part of a science fiction domain rather than an everyday achievable possibility that, for example, a target of a 15% reduction by the year 2015 would be. At a summer school organized by the UK's Energy Research Centre (UKERC), and at other interdisciplinary events on energy research that I attended during my research, directors and managers working for UKERC or for other governmental bodies concerned with energy research approached the 2050 targets as unachievable, adding in their public presentations with an ironic smile or manic grin that "fortunately I won't be here by then" (from my field notes). As they estimated their biological disappearance to occur before 2050, the speakers did not consider themselves responsible for achieving the targets; instead, the "future" in itself was expected to provide some fortuitous solutions.[2] If this is how people in charge, who make a living by formulating funding calls for energy research projects, see their role in reducing energy demand, then how could one translate this environmental concern to everyday people? While the 2050 targets are accompanied by an interim target to reduce emissions by 34% in 2020 and by "carbon budgets" covering periods of five years, these intermediate targets are not widely and publically discussed because the 2050 targets frame the general debate.

One could argue that a linear timeframe in which the end point is beyond the biological mortality of most of the members of the actual government and parliament, and whose segments—the carbon budgets—are made by set numbers rather than by types of proposed actions, might not be the best form of temporality for laypeople to act within. This is not a critique directed towards the original proponents and signatories of the Climate Change Act, but just an observation about the incommensurability of the targets and the timeframe proposed by this document in relation to the everyday domestic temporalities and understandings of agency that laypeople enact.

In this context, I see my position as a researcher working for an applied interdisciplinary project concerned with reducing energy demand in a given temporal framework as responding to a double bind. First, in my relationships with the research participants, by being part of the LEEDR project, I am regarded as a representative of the linear temporality of environmental concerns that is set by the Climate Change Act. Secondly, as an anthropologist who completed long-term ethnographic fieldwork, I wish to give voice to my participants and to describe, conceptualize, and communicate the models of time and the forms of ordinary agency that they employ in their everyday lives to wider institutions and structures. This conflict of temporalities might be what made my findings possible in the first place because the people who took part in my research welcomed me into the nitty-gritty of their everyday lives and talked about and enacted ordinary agencies in my presence while I was, arguably, a representative of the "neoliberal agency" model (Gershon 2011) assumed by the linear timeframe of the 2050 targets. The fact that the participants had a conversation with me—involving propositional and tacit knowledge—about the time modes of spontaneity and anticipation could, therefore, be seen as a response to the linear timeframe that I was representing. I see the act of accommodating these vernacular models of time and agency in my research not as a question of translation to "higher" institutions and structures, but rather as a question of mediation.

Therefore, in order to reconcile the linear timeframe expressed by the Climate Change Act—together with the applied project that my research was part of—and the temporalities of everyday life employed by the people who took part in my research, I suggest that more attention and effort can be put into the way in which the goals of reducing the country's carbon emissions are formulated. It can be argued that actions of spontaneity and anticipation, like other human actions, involve energy consumption, for example, by using digital media to find out the answer to a question of momentous concern and leaving the light on for a family member that is expected to return home. But what they also do is generate agency; they make people feel that they can contribute to changing the world. They are, in Foucault's (1990, 2000) terms, "practices of freedom." Therefore, it is exactly these everyday actions that should be encouraged, and not supressed, by "behaviour change" interventions and strategies developed in relation to the domestic sustainability agenda. By starting from a "homes as sites for action" premise and acknowledging everyday domestic moments of interruption as forms of political action (Hobson 2011), a collaborative approach to sustainability could emerge. The recognition and adoption of a more complex model of time and of agency in the formulation of the carbon emission reduction goals could be translated into a variety of propositions that would encourage public debate and collective action.

By taking inspiration from the time mode of anticipation, schedules and slots for energy saving could be proposed as cyclical occurrences or in relation to events that are recognized and celebrated widely, such as Christmas. For example, December could be recommended as a month when people should switch off all their energy-consuming appliances every evening ten minutes earlier than normal in preparation for the higher energy consumption that might occur during the Christmas holiday. To commit to and follow this proposition would be a form of marking time itself by counting down the days before Christmas with an action that would be as visible as using an Advent calendar. Besides celebratory events, anticipation in relation to a cyclical time model could be engaged by introducing other energy-saving suggestions related to domestic everyday preoccupations, such as the challenge of not using the tumble dryer every second weekend of the month regardless of how the weather is or switching the heating off earlier on Mondays as a post-weekend action aimed at making up for the supposedly higher levels of energy consumption that were reached on Saturday and Sunday.

In order to address and make positive use of the time mode of spontaneity, a competition to save energy through hacking could be proposed. This would involve building low energy-consuming devices from scratch, or combining various devices in a creative way that would lead to saving energy. As discussed in the previous chapters, the people who took part in my research were already used to finding out ways of combining technologies in new assemblages that showed their creativity and individuality. A good example of everyday creativity in designing assemblages of technologies is in Iris and Steve's recounting of the way in which they created a Christmas hub in their kitchen by playing a video of a crackling fire on the TV while broadcasting an online Christmas jukebox radio station through speakers connected to their tablet. This form of creativity that involves spontaneous actions could be encouraged and directed towards the goal of saving energy.

I have given a few examples of calls to collective action that would fit into existent domestic temporalities. These types of programs could provide the opportunity for the people who would respond and act to feel empowered and responsible and take further action in relation to environmental or other types of social concerns. With these ideas, I want to show that the incongruity between the models of time assumed in environmental policy and the ones that people enact in the realm of everyday life are not unresolvable; rather, it is an issue that needs to be observed and tackled rather than ignored. By developing a more elaborate approach in environmental policy to time that takes into consideration forms of ordinary agency and the temporal modalities of the everyday, this incongruity could be transformed into complexity, and the conflict experienced by a double-bind researcher could reach a creative resolution.

NOTES

1. See also Middlemiss's (2010) discussion of the ways in which individual responsibility is always linked to the acts of "structural players" like governments or corporations.
2. For some interesting insights into the processes of knowledge production and decision-making related to energy systems in the US, see Nader (1981, 1996, 2010).

REFERENCES

Adam, Barbara. 1995. *Timewatch: The Social Analysis of Time.* London: Polity Press.
Albert, B. 2002. "'Temporal Diversity': A Note on the 9th Tutzing Time Ecology Conference." *Time & Society* 11 (1): 89–104.
Barry, Andrew, Georgina Born, and Gisa Weszkalnys. 2008. "Logics of Interdisciplinarity." *Economy and Society* 37 (1): 20–49.

Beck, Ulrich, and E. Beck-Gernstein. 2002. *Individualization: Institutional Individualism and Its Social and Political Implications.* London: Sage.

Cordner, Christopher. 2008. "Foucault, Ethical Self-Concern and the Other." *Philosophia* 36 (4): 593–609.

Douglas, Mary. 1991. "The Idea of a Home : A Kind of Space." *Social Research* 58 (1): 287–307.

Ellsworth-Krebs, Katherine, Louise Reid, and Colin J. Hunter. 2015. "Home-Ing in on Domestic Energy Research: 'House,' 'Home,' and the Importance of Ontology." *Energy Research & Social Science* 6. Elsevier Ltd: 100–108.

Foucault, Michel. 1990. *The History of Sexuality,* vol. 2. London: Penguin Books.

Foucault, Michel. 2000. *Ethics: Subjectiviy and Truth (Essential Works of Foucault 1954–1984).* London: Penguin Books.

Fraser, Nancy. 1990. "Rethinking the Public Sphere : A Contribution to the Critique of Actually Existing Democracy." *Social Text* 25 (26): 56–80.

Geissler, K. A. 2002. "A Culture of Temporal Diversity." *Time and Society* 11 (1): 131–140.

Gershon, Ilana. 2011. "Neoliberal Agency." *Current Anthropology* 52 (4): 537–555.

Giddens, Anthony. 1994. *Beyond Left and Right.* Cambridge: Polity Press.

Greenhouse, Carol J. 1996. *A Moment's Notice: Time Politics Across Cultures.* New York: Cornell University Press.

Hargreaves, Tom, Michael Nye, and Jacquelin Burgess. 2010. "Making Energy Visible: A Qualitative Field Study of How Householders Interact with Feedback from Smart Energy Monitors." *Energy Policy* 38 (10). Elsevier: 6111–6119.

Hawkins, Gay. 2011. "Interrogating the Household as a Field of Sustainability." In *Material Geographies of Household Sustainability,* edited by Ruth Lane and Andrew Gorman-Murray, 69–72. Farnham, UK: Ashgate.

Henning, Annette. 2005. "Climate Change and Energy Use: The Role for Anthropological Research." *Anthropology Today* 21 (3): 8–12.

Hobson, Kersty. 2003. "Thinking Habits into Action : The Role of Knowledge and Process in Questioning Household Consumption Practices." *Local Environment* 8 (1): 95–112.

Hobson, Kersty. 2011. "Environmental Politics, Green Governmentality and the Possibility of a 'Creative Grammar' for Domestic Sustainable Consumption." In *Material Geographies of Household Sustainability,* edited by Ruth Lane and Andrew Gorman-Murray, 193–210. Farnham, UK: Ashgate.

Hochschild, Arlie. 1997. *The Time Bind: When Work Becomes Home and Home Becomes Work.* New York: Metropolitan Books.

Hoffman, Steven M., and Angela High-Pippert. 2005. "Community Energy: A Social Architecture for an Alternative Energy Future." *Bulletin of Science, Technology & Society* 25 (5): 387–401.

Macfarlane, Alan. 1978. *The Origins of English Individualism: The Family, Property and Social Transition*. Oxford: Blackwell.

Macnaghten, Phil. 2003. "Embodying the Environment in Everyday Life Practices." *The Sociological Review* 51 (1): 63–84.

Middlemiss, Lucie. 2010. "Reframing Individual Responsibility for Sustainable Consumption: Lessons from Environmental Justice and Ecological Citizenship." *Environmental Values* 19 (2): 147–167.

Miyazaki, Hirokazu. 2004. *The Method of Hope: Anthropology, Philosophy, and Fijian Knowledge*. Stanford: Stanford University Press.

Moore, Henrietta L. 2011. *Still Life: Hopes, Desires and Satisfactions*. Cambridge: Polity Press.

Morgan, David H.J. 1996. *Family Connections: An Introduction to Family Studies*. Cambridge: Polity Press.

Nader, Laura. 1981. "Barriers to Thinking New about Energy." *Physics Today* 34 (9): 99–104.

Nader, Laura. 1996. "The Three-Cornered Constellation: Magic, Science and Religion Revisited." In *Naked Science: Anthropological Inquiry into Boundaries, Power, and Knowledge*, edited by Laura Nader. New York: Routledge.

Nader, Laura. 2010. *The Energy Reader*. Oxford: Wiley-Blackwell.

Norgaard, Kari Marie. 2011. *Living in Denial: Climate Change, Emotions, and Everyday Life*. Cambridge, MA: The MIT Press.

Pepper, David. 1991. *Communes and the Green Vision: Counterculture, Lifestyle and the New Age*. London: Green Print.

Rabinow, Paul, George E. Marcus, James D. Faubion, and Tobias Rees. 2008. *Designs for an Anthropology of the Contemporary*. Durham, NC: Duke University Press.

Rutland, Ted, and Alex Aylett. 2008. "The Work of Policy: Actor Networks, Governmentality, and Local Action on Climate Change in Portland, Oregon." *Environment and Planning D: Society and Space* 26 (4): 627–646.

Serres, Michel, and Bruno Latour. 1995. *Conversations on Science, Culture, and Time*. Ann Arbor: The University of Michigan Press.

Sillitoe, Paul. 2007. "Anthropologists Only Need Apply: Challenges of Applied Anthropology." *Journal of the Royal Anthropological Institute* 13 (1): 147–165.

Silva, Elizabeth Bortolaia, and Carol Smart. 1999. "The 'New' Practices and Politics of Family Life." In *The New Family?*, edited by Elizabeth Bortolaia Silva and Carol Smart. London: Sage.

Slocum, Rachel. 2004. "Consumer Citizens and the Cities for Climate Protection Campaign." *Environment and Planning A* 36 (5): 763–782.

Strathern, Marilyn. 1992. *After Nature: English Kinship in the Late Twentieth Century*. Cambridge: Cambridge University Press.

Strathern, Marilyn. 2004. *Commons and Borderlands: Working Papers on Interdisciplinarity, Accountability and the Flow of Knowledge*. Wantage: Sean Kingston Publishing.

Strathern, Marilyn. 2006. "A Community of Critics? Thoughts on New Knowledge." *Journal of the Royal Anthropological Institute* 12 (1): 191–209.

Strauss, Sarah, Stephanie Rupp, and Thomas Love. 2013. "Introduction. Powerlines: Cultures of Energy in the Twenty-First Century." In *Cultures of Energy: Power, Practices, Technologies*, edited by Sarah Strauss, Stephanie Rupp, and Thomas Love. Walnut Creek, CA: Left Coast Press.

Taylor, Charles. 1979. "Action as Expression." In *Intention and Intentionality: Essays in Honour of G.E.M. Anscombe*, edited by Cora Diamond and Jenny Teichmann, 73–89. Brighton, UK: Harvester Press.

Urry, John. 2011. *Climate Change and Society*. Cambridge: Polity Press.

Yanagisako, Sylvia Junko. 1979. "Family and Household." *Annual Review of Anthropology* 8. Routledge: 161–205.

Yanagisako, Sylvia Junko, and Carol Delaney. 1995. "Naturalizing Power." In *Naturalizing Power: Essays in Feminist Cultural Analysis*, edited by Sylvia Yanagisako and Carol Delaney, 1–23. New York: Routledge.

INDEX

© The Editor(s) (if applicable) and The Author(s) 2016
R. Moroşanu, *An Ethnography of Household Energy Demand
in the UK*, DOI 10.1057/978-1-137-59341-2